OXFORD STATISTICAL SCIENCE SERIES

OXFORD STATISTICAL SCIENCE SERIES

---

1. A.C. Atkinson: *Plots, transformations, and regression*
2. M. Stone: *Coordinate-free multivariable statistics*
3. W.J. Krzanowski: *Principles of multivariate analysis: a user's perspective*
4. M. Aitkin, D. Anderson, B. Francis, and J. Hinde: *Statistical modelling in GLIM*
5. Peter J. Diggle: *Time series: a biostatistical introduction*
6. Howell Tong: *Non-linear time series: a dynamical approach*
7. V.P. Godambe: *Estimating functions*
8. A.C. Atkinson and A.N. Donev: *Optimum experimental designs*
9. B.R. Bhat and I.V. Basawa: *Queuing and related models*
10. J.K. Lindsey: *Models for repeated measurements*
11. N.T. Longford: *Random coefficient models*
12. P.J. Brown: *Measurement, regression, and calibration*
13. Peter J. Diggle, Kung-Yee Liang, and Scott L. Zeger: *Analysis of longitudinal data*
14. J.I. Ansell and M.J. Phillips: *Practical methods for reliability data analysis*
15. J.K. Lindsey: *Modelling frequency and count data*

Modelling Frequency and Count Data

J.K. LINDSEY

Department of Biostatistics
Limburgs Universitair Centrum
Diepenbeek, Belgium
jlindsey@luc.ac.be

Department of Social Sciences
University of Liege
Liege, Belgium

CLARENDON PRESS • OXFORD

1995

Oxford University Press, Walton Street, Oxford OX2 6DP

Oxford New York
Athens Auckland Bangkok Bombay
Calcutta Cape Town Dar es Salaam Delhi
Florence Hong Kong Istanbul Karachi
Kuala Lumpur Madras Madrid Melbourne
Mexico City Nairobi Paris Singapore
Taipei Tokyo Toronto
and associated companies in
Berlin Ibadan

Oxford is a trade mark of Oxford University Press

Published in the United States by
Oxford University Press Inc., New York

A catalogue record for this book is available from the British Library

Library of Congress Cataloging in Publication Data
(Data available)

ISBN 0 19 852331 9

Typeset by the author using LaTeX

Printed in Great Britain by
Bookcraft (Bath) Ltd
Midsomer Norton, Avon

Preface

The present text originated from teaching a third-year categorical data course to non-mathematics majors. It was originally planned to be a second edition of Lindsey (1989), but the evolution of statistics and software dictated that it become something else.

In recent years, it has become increasingly obvious that the basis of applied statistics is not the normal distribution, but the multinomial and Poisson distributions. Let us leave the former to those statisticians working on the asymptote. Applied statistics is essentially the study of how histograms change their shape under different conditions (Lindsey, 1995a). This is a general form of regression problem, where responses depend on the conditions under which they are produced. Parametric statistics can thus be viewed as a series of special cases of categorical data analysis. Unfortunately, this is not reflected in the way statistics is currently presented to students.

Categorical data analysis is concerned with the study of events. Models can be constructed to describe the structure in such data, that is, to describe their data generating mechanism. Events may be observed to occur to independent individuals, yielding what I shall call frequencies of (independent) events. Or the same types of event(s) may repeatedly occur to the same individuals, yielding counts of (dependent) events. This important distinction has not been sufficiently emphasized in the categorical data literature. Thus, the two main themes underlying the presentation will be the use of models to find structure in data and the distinction between frequencies and counts.

It is now well known that log linear and logistic models are a special case of generalized linear models (Lindsey, 1973). Much less well known is the fact that generalized linear models are, in turn, a special case of log linear models for any empirically observable data, as are 'continuous' time proportional hazards models. This will be a third theme of the book. As already mentioned, we shall be primarily interested in regression-type problems; we shall spend little time on purely multivariate data composed uniquely of several different types of responses.

Given the importance of the analysis of categorical data using log linear models, and their connection with generalized linear models, this book should be of interest to anyone, in any field, concerned with such applications. It should be suitable as a manual for applied statistics courses covering the subject of categorical data. To this end, examples and exercises have been selected from a wide variety of different fields. The idea has been to cover what can be done with generalized linear modelling software. On the other hand, discussion of the inference aspects has been kept to a minimum. Unfortunately, this has meant that important topics, such as exact tests for sparse data tables, have had to be ignored. For a more complete exposition of the statistical properties of log linear/logistic models, many excellent books already exist, of which Agresti (1990) and Collett (1991) particularly stand out.

I assume that the reader has already had an introductory course in statistics, and specifically some previous knowledge of log linear/logistic models (which, by now, should be a basic element of any such introductory course!) and of generalized linear models. My students in this course learn their basic statistics from Lindsey (1995a). Little mathematical sophistication is required for most of the text. However, several chapters (6, 7, and 10) contain considerably more complex material than the others, and will be difficult for nonmathematics students.

I also assume that the reader has access to software permitting the fitting of generalized linear models. At present, the best known include GENSTAT, GLIM4, LISP-STAT, SAS, and S-Plus. Software which uses the Wilkinson and Rogers (1973) notation (GENSTAT, GLIM, S-Plus) will make replication of the examples especially easy, because that notation will be used in this book for specifying the more complex models.

The results for all of the examples, including the graphics, were produced with GLIM4. However, several of the models used cannot be fitted as standard generalized linear models. Special GLIM4 code for these is supplied in an Appendix. Hopefully, this will provide indications as to how such models could be fitted with other generalized linear modelling software.

I would like to thank my students who have worked through this course over the past years and supplied invaluable reactions and comments, the developers of GLIM without whom this work would not have been possible, Philippe Lambert, Mark Becker, and Joyce Snell for many useful comments, and R. Doutrelepont and C. Laurent, who supplied the data for two of the examples. Much of the preliminary work for the present text was done while I was Visiting Senior Research Fellow at the Centre for Applied Statistics at Lancaster University.

I also thank all of the contributors of data sets; they are individually cited when each table is first presented.

Diepenbeek J.K.L.
May, 1994

Contents

Part I

Frequency data

1
One-way frequency tables

This is a book on modelling categorical data, often called discrete data analysis. Such observations involve enumerating one or more types of event. A variable can be defined by an exhaustive and mutually exclusive list of these types. If these events all occur independently on different units or individuals, we call the enumerated results *frequencies*. This is the subject of the first part of the book. On the other hand, if several events from the same variable are observed on the same unit, they will be called *counts*. In the latter case, we would expect some kind of dependency among the repeating events. The second part of the book looks at this type of event observation. Thus, if we study the number of children in families, these values (0, 1, 2, ...) would be counts within each family unit, while the number of independent families with, say, k children would be a frequency.

Counts may involve one or more categories of event; having a child is an event for the family. On the other hand, frequencies are only of interest when there are at least two categories, although these may be 'have an event of interest' or not. Thus, both are aspects of categorical data analysis. Some of the models may be similar for the two types of data. Nevertheless, models for count data must usually take into account the specific problem of dependence among events. For the rest of this part, we shall concentrate on frequency data.

1.1 Models for events

If we are interested in events, we are also interested in the probability of their occurrence. Let π_i represent the probability of the type of event i ($i = 1, \ldots, I$). Such probabilities must be positive values and they must sum to unity. These are the minimal requirements for a suitable model. The simplest way, although not the only one, to ensure that the first condition is fulfilled is to construct models in terms of the (natural) logarithms of the probabilities, $\log(\pi_i)$. As we shall see, the second condition will be fulfilled by fixing the total number of observations at its observed value.

The simplest frequency tables concern a single variable and show the frequencies with which the various categories of that variable have been observed. For the moment, we shall be interested in tables where the variable may be nominal, ordinal, integral (together often called discrete), or continuous, but where the only assumption is that we have categories of different independent events. Thus, a multinomial distribution, which multiplies independent probabilities of events, should be appropriate.

In those cases where the categories refer to some continuous measure, such as income, length of employment, etc., or are themselves counts, such as numbers of accidents per individual, number of children per family, etc., special cases of the multinomial distribution with specific functional forms, called probability distributions, can often be fitted to the data. We shall look at modelling these in the categorical data context in Chapter 6.

Most software for generalized linear models does not handle the multinomial distribution directly. Nevertheless, we can demonstrate very simply that models based on this distribution,

$$\Pr(n_1, \ldots, n_I) = \binom{n_\bullet}{n_1 \cdots n_I} \pi_1^{n_1} \cdots \pi_I^{n_I}$$

where $n_1 \cdots n_I$ are the frequencies, with total $n_\bullet$, and $\pi_1 \cdots \pi_I$ the corresponding probabilities, can equivalently be analysed by models which such software does treat, that is, those based on the Poisson distribution

$$\Pr(n_i) = \frac{\mathrm{e}^{-\nu_i} \nu_i^{n_i}}{n_i!}$$

for each category, if we condition on the total number of observations. Here, ν_i is the theoretical average *number* of events of type i, while π_i is the theoretical *proportion* of events of that type. Let us recall two points of probability theory. First, the conditional probability for event A given event B is defined by

$$\Pr(A|B) = \frac{\Pr(A \text{ and } B)}{\Pr(B)}$$

Second, if a set of frequencies, $n_1 \cdots n_I$, has a Poisson distribution with means $\nu_1 \cdots \nu_I$, then their sum, $n_\bullet$, also has a Poisson distribution with mean, $\nu_\bullet$, the sum of the individual means.

We are now in a position to demonstrate the relationship between the multinomial and the conditional Poisson distributions:

$$\begin{aligned}
\Pr(n_1, \ldots, n_I | n_\bullet) &= \frac{\prod_{i=1}^{I} \mathrm{e}^{-\nu_i} \nu_i^{n_i} / n_i!}{\mathrm{e}^{-\nu_\bullet} \nu_\bullet^{n_\bullet} / n_\bullet!} \\
&= \frac{n_\bullet! \mathrm{e}^{-\nu_\bullet} \prod_{i=1}^{I} \nu_i^{n_i}}{\prod_{i=1}^{I} n_i! \mathrm{e}^{-\nu_\bullet} \nu_\bullet^{n_\bullet}}
\end{aligned}$$

Table 1.1. Subjects reporting one stressful event. (Haberman, 1978, p. 3)

Months Before	1	2	3	4	5	6	7	8	9
Number	15	11	14	17	5	11	10	4	8
Months Before	10	11	12	13	14	15	16	17	18
Number	10	7	9	11	3	6	1	1	4

$$= \binom{n_\bullet}{n_1 \cdots n_I} \prod_{i=1}^{I} \left(\frac{\nu_i}{\nu_\bullet}\right)^{n_i}$$

which is the multinomial distribution with $\pi_i = \nu_i/\nu_\bullet$, as might be expected, and the two distributions are identical. In this way, we are fixing $\nu_\bullet = n_\bullet$ as known, and ensuring that the sum of the probabilities is unity.

1.2 Time trend model

Before going further, we shall consider our first example; the data given in Table 1.1 were collected to study the relationship between life stresses and illnesses. One randomly chosen member of each randomly chosen household in a sample from Oakland, California, U.S.A., was interviewed. In a list of 41 events, respondents were asked to note which had occurred within the last 18 months. The results given are for those recalling only one such stressful event.

Our variable classifies subjects according to the number of months prior to an interview that they remember a stressful event. We wish to determine if the probability of an event is the same through all 18 months, which might be reasonable if events occurred at a constant rate and if recall was equally good for all times in the past. If we look at Table 1.1, we see immediately that the frequency of events remembered seems, in fact, to decrease with time. This is confirmed if we plot the data, as in Figure 1.1, ignoring, for the moment, the curved line.

To start any analysis, computer software handling generalized linear models will require certain basic information. We must specify the names of the variables in the data set, the format of the data, which variable is the response (to be explained below), what probability distribution to use, and what linear model to fit. A linear model is specified by the list of variables which it contains, each separated by an operator (+, ·, *). The + has the usual model meaning of adding terms together, while the · and * signify interactions. Thus,

$$\mathtt{A} + \mathtt{B} + \mathtt{A} \cdot \mathtt{B} \equiv \mathtt{A} * \mathtt{B}$$

specifies the main effects of A and B and their interaction.

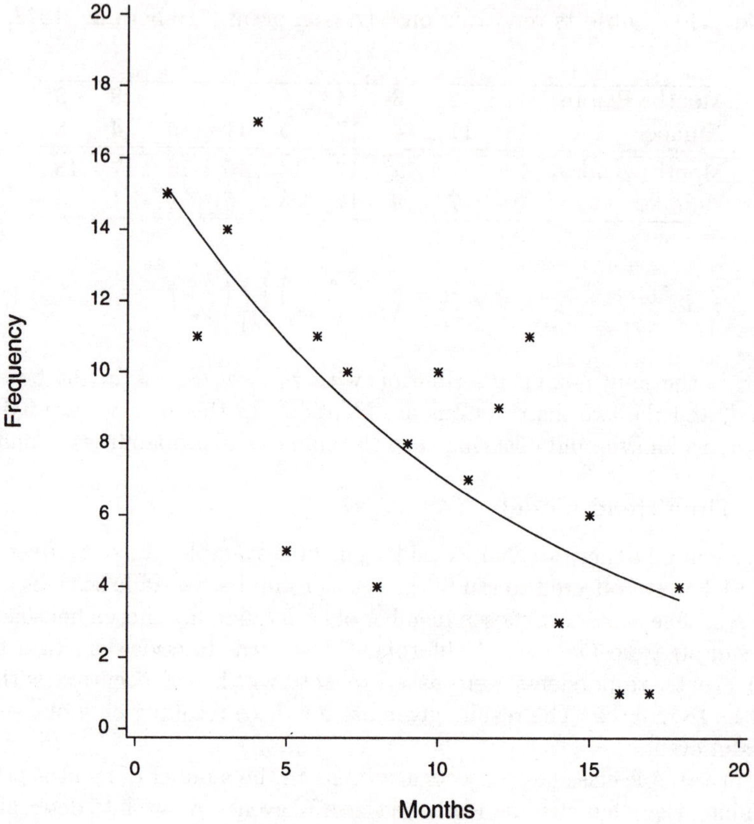

Fig. 1.1. Observed frequencies and fitted exponential decay model of Equation (1.3) for the recall data in Table 1.1.

For these data, we require a Poisson distribution, with the frequencies as 'response' variable, and fit a general mean or intercept, called a null model. The Poisson distribution in such models is fitted by successive approximations or iterations. One piece of information which the software will then provide is the deviance or log likelihood ratio, where the former is defined as -2 times the latter. This compares the model being fitted to a saturated model with a different parameter value for each observation. The larger the deviance, the further is the fitted model from an exact fit to the data. However, closeness to the data also depends on the number of parameters, say k, in the model or on the degrees of freedom (d.f.), $n-k$, the number of parameters that could still be added to the model. A balance between the size of the deviance and the number of parameters is often difficult to judge. One way to do this is by the Akaike information criterion (AIC), which adds $2k$ to the deviance. A model with a smaller AIC is preferable; it should be less than twice the number of frequencies in the table (the AIC for the saturated model). We shall give this in parentheses after each deviance in the text.

When we fit our equiprobability model, with constant mean for all months, we find the deviance to be 50.84 (52.84) with 17 d.f., which reveals a considerable lack of fit. The probability of recall appears not to be the same for all of the months.

We must now elaborate on the model which we have just fitted. This is what is commonly called a log linear model, because it is linear in the logarithms of the means. Specifically, we have fitted a common mean, or equivalently a constant probability, to all of the categories:

$$\begin{aligned}\log(\nu_i) &= \log(n_\bullet \pi_i) \\ &= \mu \qquad \text{for all } i \end{aligned} \tag{1.1}$$

The maximum likelihood estimate, $\hat{\mu} = 2.100$ (with standard error 0.0825), will be the second important piece of information provided by the software.

In this model, all observations are estimated by the same fitted value, 8.167, because our model only contains the mean. This can be related to our probabilities:

$$\begin{aligned}\tilde{\pi}_i &= \frac{\hat{\nu}}{n_\bullet} \\ &= \frac{8.167}{147} \\ &= \frac{1}{18}\end{aligned}$$

The $\tilde{}$ on π_i indicates that this is the best estimate from the data for the given model, but not necessarily the best overall estimate of that parameter from the data.

We should note that all of our analysis up until now applies to any set of frequencies whether structured or not. Our classification variable could have been nominal because we have not yet used the ordering of the months in this constant probability model.

The Poisson residuals, which are differences between observed and fitted values divided (standardized) by their estimated standard errors, i.e. $\sqrt{\hat{\mu}_i}$, can also be displayed, as shown in Table 1.2. Let us examine these more closely. We see that the first four are positive and the last five are negative, indicating that the probability of recalling an event is more than average in recent months and less than average over the longer time period, which is as one might expect.

We may now introduce this ordering so as to study the observed decrease

Table 1.2. Fitted values and residuals from the equiprobability model of Equation (1.1) for the recall data in Table 1.1.

Observed	Fitted	Residual
15	8.167	2.391
11	8.167	0.991
14	8.167	2.041
17	8.167	3.091
5	8.167	−1.108
11	8.167	0.991
10	8.167	0.642
4	8.167	−1.458
8	8.167	−0.058
10	8.167	0.642
7	8.167	−0.408
9	8.167	0.292
11	8.167	0.991
3	8.167	−1.808
6	8.167	−0.758
1	8.167	−2.508
1	8.167	−2.508
4	8.167	−1.458

in number of events remembered. Let us suppose that the probability of remembering an event diminishes in the same proportion between any two consecutive months:

$$\frac{\pi_i}{\pi_{i-1}} = \phi$$

a constant for all i. Then, by recursion,

$$\frac{\pi_i}{\pi_1} = \phi^{i-1} \tag{1.2}$$

and

$$\log\left(\frac{\pi_i}{\pi_1}\right) = (i-1)\log(\phi)$$

but

$$\pi_i = \frac{\nu_i}{n_\bullet}$$

so that

$$\log\left(\frac{\pi_i}{\pi_1}\right) = \log\left(\frac{\nu_i}{\nu_1}\right)$$

and

$$\begin{aligned}\log(\nu_i) &= \log(\nu_1) + (i-1)\log(\phi) \\ &= \log\left(\frac{\nu_1}{\phi}\right) + i\log(\phi)\end{aligned}$$

which may be rewritten

$$\log(\nu_i) = \beta_0 + \beta_1 i$$

where

$$\beta_0 = \log\left(\frac{\nu_1}{\phi}\right)$$

and

$$\beta_1 = \log(\phi)$$

This is a log linear time trend or Poisson regression model, a special case of (log) linear regression. To perform the analysis for this model, we must construct a variable for months. This can be a vector, `MONTH`, of length 18 filled with the integers from 1 to 18. We then fit the model, by adding `MONTH`, and examine the required information, as in the previous case.

The new deviance is 24.57 (28.57) with 16 d.f. (The saturated model, with 18 parameters, has an AIC of 36.) By adding one parameter, and losing one degree of freedom, we have reduced our deviance by 26.27 and the AIC by 24.27. We have a model which fits more closely to the data, while being only slightly more complex. This indicates that we must reject our equiprobability model in favour of the one for constant reduction in probability. The remaining deviance of 24.57, with 16 d.f., indicates that additional, non-linear effects, i.e. a non-constant reduction in probability, may not need to be taken into account.

Our estimated model is now

$$\log(\tilde{\nu}_i) = 2.803 - 0.0838i \tag{1.3}$$

The negative value of β_1, the slope parameter for months, indicates decrease in probability with time elapsed. Because $\beta_1 = \log(\phi)$, $\phi = e^{\beta_1}$, so that $\phi = 0.920$, the proportional decline in probability per month. If we rewrite our model of Equation (1.2) in terms of β_1, we have

$$\pi_i = \pi_1\phi^{i-1}$$

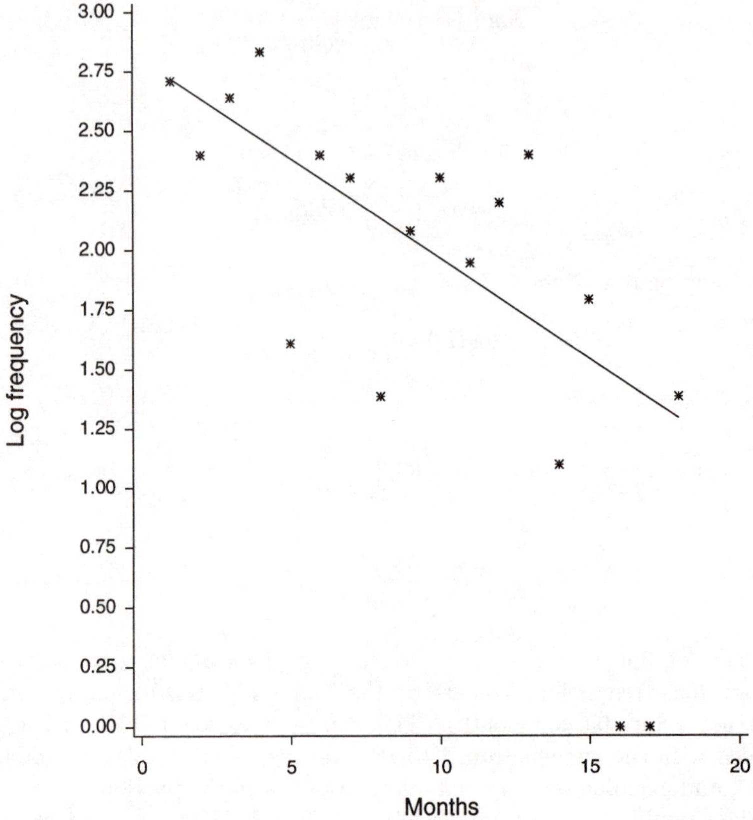

Fig. 1.2. The fitted exponential decay model of Equation (1.3) plotted for the logarithm of the mean for the recall data in Table 1.1.

$$= \pi_1 e^{\beta_1(i-1)}$$

which is a model of exponential decay. If β_1 were positive, it would be a model of exponential growth.

It is now useful to look at the plot of our model; we can plot fitted values against the month as the line in Figure 1.1. We see the form of the exponential decay in the curved line. If we take logarithms of the observed and fitted values and plot them, we obtain Figure 1.2, which is our linear model represented by the straight line, surrounded by the observed points.

We have already seen the usefulness of interpreting the residuals of a model. We should note, however, that inspection of residuals only proves useful when we have a reasonable number of degrees of freedom. As the degrees of freedom approach zero, the model must necessarily represent the

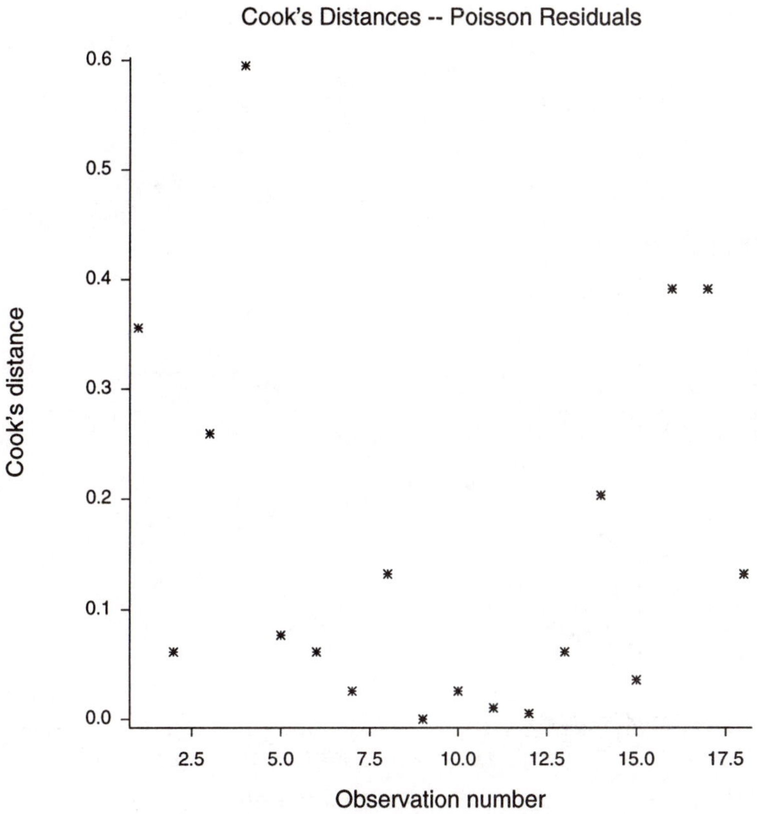

Fig. 1.3. Cook's distances from the constant probability model of Equation (1.1) for the recall data in Table 1.1.

data more closely and the residuals cannot vary very much from zero. The task of inspecting residuals is also made easier if we plot them.

The ordered standardized residuals (when further corrected by the diagonal elements of what is known as the hat matrix) have expected values which can be approximately represented by a normal distribution. This yields a normal probability (Q–Q) plot. A second useful technique is to study the influence of each observation by seeing how omitting it changes the parameter estimates. The global effect on all parameter estimates is given by a plot of Cook's distances. (Pregibon, 1982 calls this a score test coefficient of sensitivity; see also Gilchrist, 1981 and 1982a.) Both plots can be produced by the software for each model fitted. (Certain software packages are limited to only producing such plots for the model fitted immediately previously.)

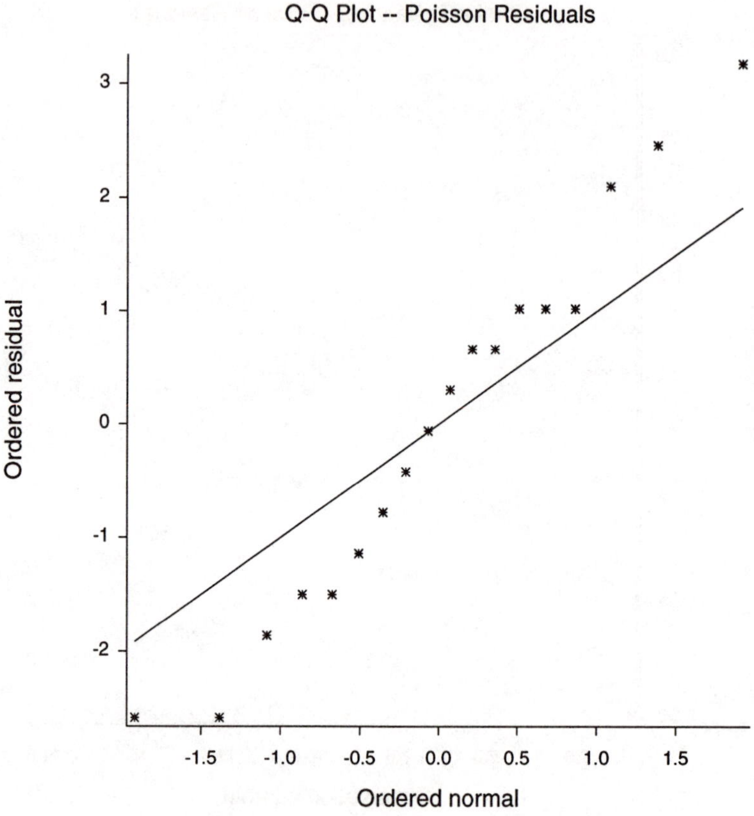

Fig. 1.4. Normal probability plot from the constant probability model of Equation (1.1) for the recall data in Table 1.1.

The plots for the constant probability model of Equation (1.1) are given in Figures 1.3 and 1.4. Those for the time trend model of Equation (1.3) are given in Figures 1.5 and 1.6.

Residual plots are primarily useful in searching for systematic patterns, which indicate structure in the data not taken into account by the model. Cook's distances are shown as a plot of the ordered individual observations as against a modified residual. The model fits those observations with large values of this statistic least well. Values smaller than one are usually of little concern. For the constant probability model, we see, in Figure 1.3, that the first and last few observations fit less well, as we have already noted above. Although none are excessively large, the pattern is informative. For the time trend model in Figure 1.5, no obvious pattern is observable and all values are much smaller than for the simpler model. The 13th month

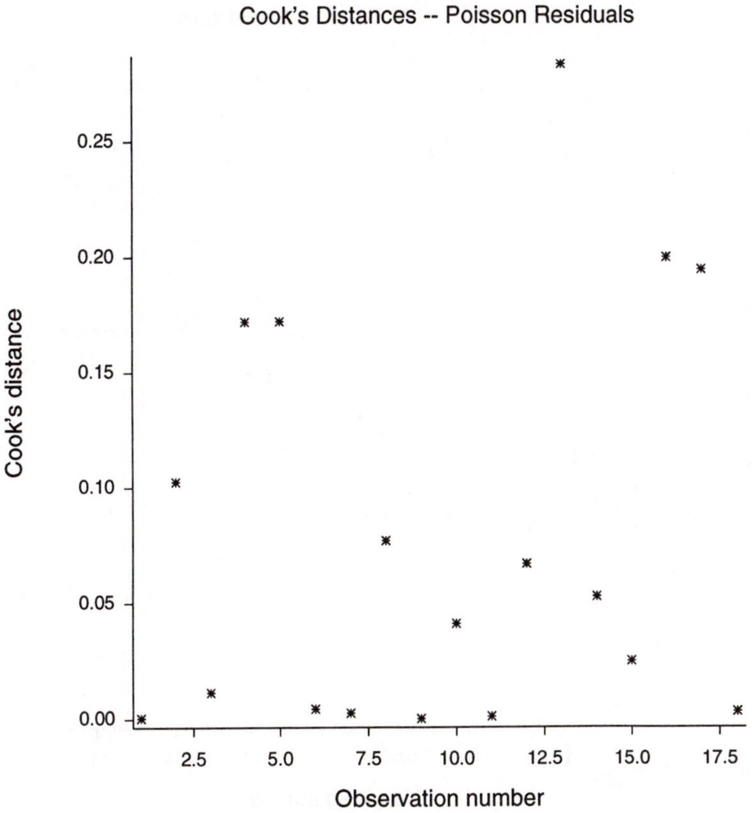

Fig. 1.5. Cook's distances from the exponential decay model of Equation (1.3) for the recall data in Table 1.1.

fits least well; from the list of residuals (not shown), we can see that 11 recalls are recorded while only 5.6 are predicted.

If a model is acceptable, the normal probability plot should be close to a 45 degree straight line. A steeper slope indicates overall lack of fit of the model, i.e. that it is not close enough to the data, while a shallower slope reveals overfitting. If the points do not lie on a straight line, the model may not be accounting for all the structure in the data. We observe, in Figure 1.4, that the residual plot for our constant probability model has a slope greater than this, indicating the lack of fit, whereas that in Figure 1.6 for the log linear trend model has about the required slope, although it is not very straight. This may be, in part, due to the tendency to remember events about one year before, especially 13 months here. We shall gain experience in interpreting such diagnostic plots as we proceed through the

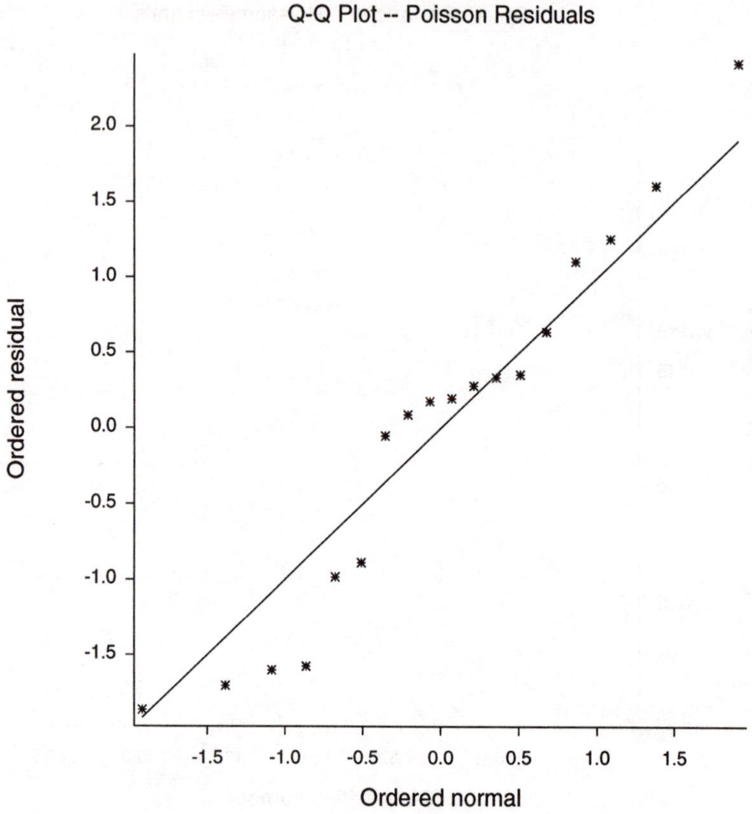

Fig. 1.6. Normal probability plot from the exponential decay model of Equation (1.3) for the recall data in Table 1.1.

examples in the book.

We have now provisionally completed the study of our first example, although we shall come back to it in Chapter 6.

1.3 Symmetry model

For our second example, with data given in Table 1.3, we shall look at how subjects self-classify themselves into four social classes: lower, working, middle, or upper class.

Study of the table shows that many fewer people have chosen the two extreme categories than the central ones. We may ask if this aversion to the extremes is symmetrical for the top and the bottom classes. Thus, we are interested in determining if the table is symmetric, i.e. if $\pi_1 = \pi_4$ and $\pi_2 = \pi_3$. This can be translated into a log linear model in the following

Table 1.3. Self-classification of individuals by social class. (Haberman, 1978, p. 24)

Lower	Working	Middle	Upper
72	714	655	41

Table 1.4. Fitted values and residuals from the first two models for the self-classification data of Table 1.3.

	Equiprobability		Symmetry	
Observed	Fitted	Residual	Fitted	Residual
72	370.5	−15.508	56.5	2.062
714	370.5	17.846	684.5	1.128
655	370.5	14.780	684.5	−1.128
41	370.5	−17.118	56.5	−2.062

terms:

$$\begin{aligned} \log(\nu_i) &= \mu + \alpha \qquad i = 1, 4 \\ &= \mu - \alpha \qquad i = 2, 3 \end{aligned} \tag{1.4}$$

For the model of Equation (1.4), we require a new variable which has, say, +1 for the first and last values and −1 for the second and third values. Let us call it `CLASS`.

First we fit the equiprobability model, as in the example on recall in Section 1.2; this has a deviance of 1266.8 (1268.8) with 3 d.f. As may be expected, this model fits very badly. The residuals given in Table 1.4 indicate the parabolic form of the relationship and support our hypothesis.

Next, we add our `CLASS` variable to obtain the symmetry model. However, this model, with a deviance of 11.16 (15.16) and 2 d.f., is also to be rejected. The parameter estimate for `CLASS`, −1.247, is negative, reflecting the fact that fewer people choose the extremes (-1.247×1) than the middle (-1.247×-1). A look at the residuals, also given in Table 1.4, indicates a linear relationship, whereby more people than expected (for this model) classify themselves in the lower classes as compared to the upper classes.

This symmetry model may be thought of as a quadratic model centred on the middle of the social class scale, as in Figure 1.7. To account for the observed residual differences, it is necessary to shift the parabola, a little to the left in this case. In other words, we need to add a linear term, say `CLASSL`, to the quadratic term already in our model. We now have

$$\log(\nu_i) = \mu + \beta_1 x_{i1} + \beta_2 x_{i2}$$

where $x_{i1} = 2(i - 2.5)$ and $x_{i2} = (i - 2.5)^2 - 1.25$.

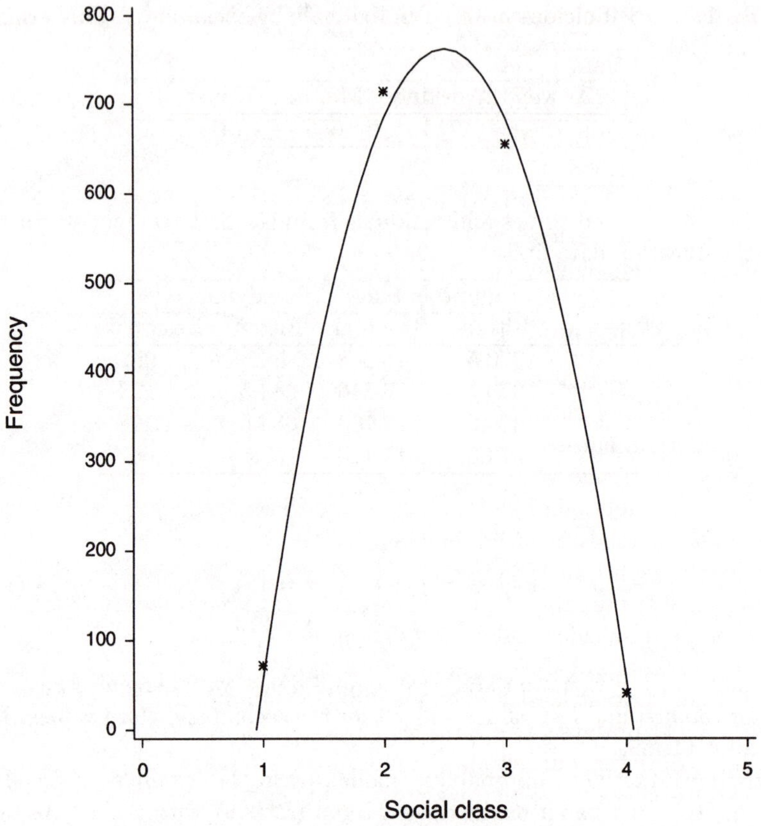

Fig. 1.7. Quadratic symmetry model for the social class self-classification data of Table 1.3.

The reader may verify that $\beta_2 x_{i2} = \pm\alpha$ for $k = 1, 2, 3, 4$, so that this parameter remains unchanged between the two models. The choice of 2.5 marks the centre of the scale for 1 to 4. The variables x_{i1} and x_{i2} are called orthogonal polynomials. They simply recode the variable as a series of vectors, linear, quadratic, cubic, ..., allowing us to study the different terms in the model as independently as possible. The sum of the product of the elements of any two vectors is zero, the definition of orthogonality. In addition, although not strictly necessary, the sum of the squares of the elements of any such vectors is usually defined to be unity. We fit this model, in generalized linear modelling software terminology (Wilkinson and Rogers, 1973), as

$$\texttt{CLASSL} + \texttt{CLASS}$$

Table 1.5. Suicides in the U.S.A., 1968. (Haberman, 1978, p. 51)

Jan.	Feb.	Mar.	Apr.	May	June
1720	1712	1924	1882	1870	1680
July	Aug.	Sept.	Oct.	Nov.	Dec.
1868	1801	1756	1760	1666	1733

and see that the deviance is now satisfactory: 1.45 (7.45) with 1 d.f. An analysis of residuals is of no use, because only 1 d.f. is left.

1.4 Periodicity models

In the second section of this chapter, we have already encountered one simple example of changes in frequency of an event with time. We shall study others in subsequent chapters. However, all such changes are not simple linear trends with time. Just as the days of the week and the seasons show a periodicity, so do many human events. One classical object of sociological study is suicide. Consider, for example, the total number of suicides per month in 1968 for the U.S.A., given in Table 1.5.

A glance at the table shows no systematic pattern: June and November have the least suicides and March the most. We may, then, first wish to test for the equiprobability of suicide throughout the year. But we must immediately face at least one minor problem: all months do not have the same number of days. The rate per day is the more pertinent parameter to study. However, our log linear model requires absolute frequencies and rates are not absolute. This may be accommodated by including a constant term for days:

$$\log(\nu_i) = \log(d_i) + \mu \tag{1.5}$$

where d_i is the number of days in the ith month. Notice that this is equivalent to studying the frequency per day, ν_i/d_i. The constant term, $\log(d_i)$, is known as an offset, because it does not involve estimation of any unknown parameters. Another similar case would be frequencies of occurrence of an event, such as deaths in various regions or cities, where the latter have different populations (see Section 7.2). Then, the offset incorporates these population sizes. Most software for generalized linear models allows the easy incorporation of such constant terms in models.

When we fit this model, the resulting deviance of 37.07 (39.07) with 11 d.f. shows that the equiprobability model must be rejected. The residuals, given in Table 1.6, seem to indicate that there are more suicides in spring and fewer in late autumn and winter. The residual plots (not shown) confirm this.

As a second step, we shall set up a model to allow for differences by

Table 1.6. Fitted values and residuals from the equiprobability model of Equation (1.5) for the U.S.A. suicide data in Table 1.5.

Observed	Fitted	Residual
1720	1810	−2.120
1712	1693	0.452
1924	1810	2.675
1882	1752	3.111
1870	1810	1.406
1680	1752	−1.716
1868	1810	1.359
1801	1810	−0.216
1756	1752	0.100
1760	1810	−1.180
1666	1752	−2.050
1733	1810	−1.814

season:

$$\log(\nu_i) = \log(d_i) + \mu + \alpha_j \qquad \begin{aligned} j &= 1 \text{ for } i = 1,2,12 \\ &= 2 \text{ for } i = 3,4,5 \\ &= 3 \text{ for } i = 6,7,8 \\ &= 4 \text{ for } i = 9,10,11 \end{aligned}$$

In this way, we allow four different probabilities of suicide, one for each season. However, as so described, our model has five parameters instead of the four required. We must add a constraint. This may be done in a number of ways, all of which are mathematically equivalent, but not all of which are as easily interpreted in a given context. One possible way is to set one α_j, say $\alpha_1 = 0$, so that the other three α_j are comparisons of these three seasons with the first. This is called the baseline constraint. It is obtained by defining what is called a factor variable, i.e. a variable with a specific number of nominal levels or categories.

For our seasonality model, the deviance is 12.60 (20.60) with 8 d.f. By the introduction of three new parameters, we obtain a large reduction of 24.47 in the deviance. And the remaining deviance is relatively small. We see confirmation that more suicides occurred in the spring and less in autumn and winter, while summer is in between: for winter (category 1), the estimate is 4.039 and for autumn (4.039 + 0.003 =) 4.042, while for spring, it is (4.039 + 0.083 =) 4.122 and for summer (4.039 + 0.024 =) 4.063. The residuals and plots (not shown) no longer indicate any clear trend, which is what we want. Note that, although residual tables and plots will not always be included in the text, due to lack of space, they should always be produced and inspected for regularities if the degrees of freedom are not too small.

An alternative, but equivalent, way of specifying constraints is to have

$$\sum_j \alpha_j = 0$$

This is known as the conventional constraint and provides us with comparisons around the mean instead of with respect to one privileged category, such as the first, as was the case in what just preceded. What is known as a design matrix must be defined. (Actually, it has already been defined in the first case as well, but, there, it is usually done automatically by the software.) This matrix is, in fact, a series of vectors, one for each parameter to be estimated. In our case, we have three parameters, the fourth being given by the sum to zero.

When we fit this model, we note immediately that the deviance is identical to the case when we used the baseline constraint. This should not be surprising because we are fitting the same model, but simply with different constraints. The parameter estimates are now (−0.0277, 0.0557, −0.0036, 0.0244). Thus, we next note that the differences between seasons are the same in the two cases. For example, in the first case, the contrast between spring and winter is given directly as 0.0834, while here it is $0.0557-(-0.0277) = 0.0834$. Thus, our interpretation does not change. In more complicated examples in the subsequent chapters, this second set of constraints will often prove invaluable as an aid to interpretation.

Here, standard errors can be of use to us. If the absolute value of the parameter estimate is considerably larger than the standard error, say twice as large, then there is a good chance that the parameter is necessary in the model. However, in contrast to a model based on the normal distribution such as classical linear regression or ANOVA, for categorical data models, the indication is only approximate and can be misleading, as we shall see in certain subsequent examples. Hence, it should always be checked by inspecting the change in deviance when the parameter is removed from the model.

If we inspect the standard errors of the estimates (not shown) for our model with baseline constraints, we notice that spring stands out as different from the other three seasons. It has the only parameter estimate which is much larger than its standard error. If we combine the three other seasons, the new deviance is 14.40 (18.40) with 10 d.f. The suicide rate is considerably higher in spring.

We appear to have here a cyclical phenomenon, but the choice of seasons as the period for the cycles may seem arbitrary for suicides. A more abstract model may be constructed using trigonometrical functions to produce harmonics describing the periodicity in the data:

$$\log(\nu_i) = \log(d_i) + \beta_0 + \beta_1 \sin[(2i-1)\pi/12] + \beta_2 \cos[(2i-1)\pi/12]$$

We note that this model has one less parameter than the seasonality one: two parameters in addition to the mean. After constructing the appropriate sine and cosine variables, we can fit the new model. The deviance is 18.71 (24.71) with 9 d.f. indicating a considerably poorer fit than the season model. The estimated model is

$$\log(\tilde{\nu}_i/d_i) = 4.067 + 0.0343 \sin[(2i-1)\pi/12] - 0.0233 \cos[(2i-1)\pi/12] \tag{1.6}$$

As in the first time trend model, a plot of observed and fitted values is also useful here. Figure 1.8 shows the observed and fitted rate of suicides per day over the 12 months. The largest deviations between observations and model on the graph occur in early summer, especially in June. This is confirmed by the residuals and plots (not shown).

A second set of harmonics may be added to the model to give

$$\begin{aligned}\log(\nu_i) = \log(d_i) &+ \beta_0 + \beta_1 \sin[(2i-1)\pi/12] + \beta_2 \cos[(2i-1)\pi/12] \\ &+\beta_3 \sin[(2i-1)\pi/6] + \beta_4 \cos[(2i-1)\pi/6]\end{aligned} \tag{1.7}$$

This model has a deviance of 7.50 (17.50) with 7 d.f., a considerably better fit than the previous one and than the seasonality models. The curve is plotted in Figure 1.9. Even this more complex model cannot meet the drastic lowering of the suicide rate in June, but the reasonably sized deviance indicates that this could be random variation. The low June rate is balanced by a high July rate, according to this model.

Both the seasonal and the harmonic models fit the data rather satisfactorily overall. The test will be to apply them to similar data for other years to see which one maintains a reasonable fit in varying circumstances (see Exercise 1.4).

1.5 Local effects

We shall now consider one final simple example with a one-way frequency table. At times, a simple model fits, and explains, only a part of the observations and the rest may be ignored in constructing that model. The latter are sometimes called outliers.

Durkheim (1897) studied the suicide rate per day, as shown in Table 1.7. In this table, more suicides occur at the beginning of the week than at the end. If we first look at the equiprobability of suicide for all days of the week, we obtain a deviance of 74.92 (76.92) with 6 d.f., not a satisfactory model. (The saturated model has an AIC of 14.)

As would be expected, we see from the first panel of residuals in Table 1.8 that the model underestimates the rate during the first four weekdays

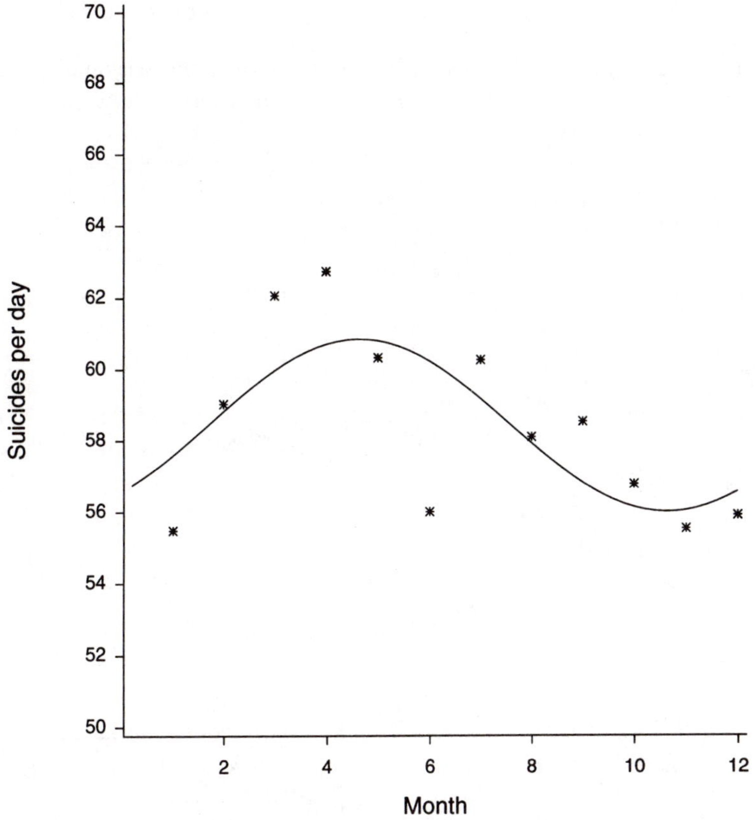

Fig. 1.8. The harmonic model of Equation (1.6) for the U.S.A. suicide data of Table 1.5.

Table 1.7. Daily suicides. (Durkheim, 1897, p. 101)

Mon.	Tues.	Wed.	Thurs.	Fri.	Sat.	Sun.
1001	1035	982	1033	905	737	894

and overestimates that for Friday and the weekend. Saturday especially stands out, with a much larger residual than any other day.

In a second model, we construct a binary variable to distinguish between these two periods of the week. Now, the residuals, also in Table 1.8, indicate that reasonable estimates appear to be given for the four weekdays, but Friday and the weekend, especially Saturday, still pose a problem. With a deviance of 23.35 (27.35) for 5 d.f., the model, although much better, is not yet acceptable.

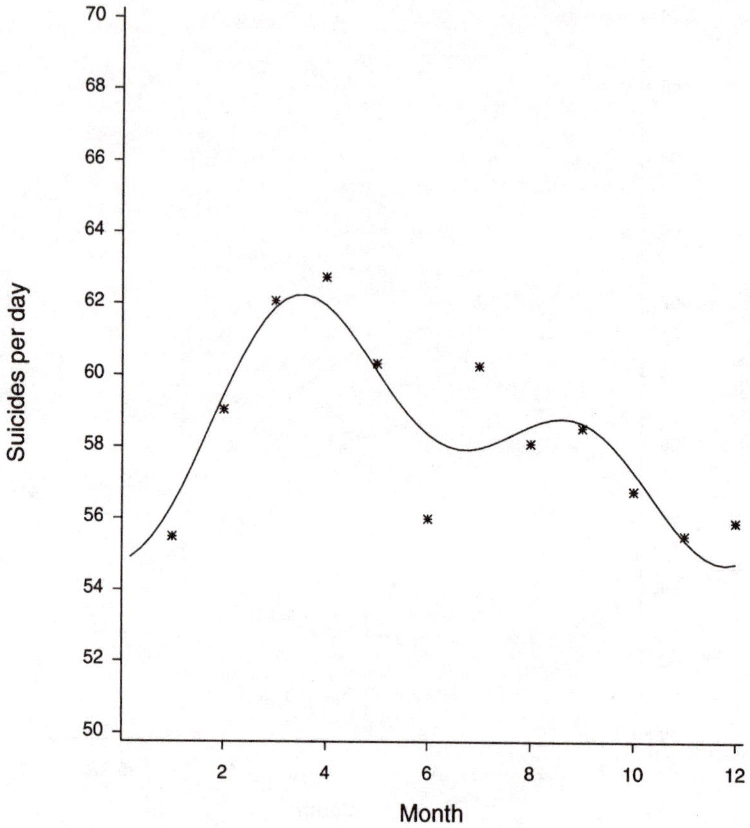

Fig. 1.9. The second harmonic model of Equation (1.7) for the U.S.A. suicide data of Table 1.5.

Let us, then, ignore completely these three days, which seem to vary among themselves, and fit an equiprobability model to the four weekdays. To do this, we can introduce a weighting factor, with unit weight for the days of interest and zero weight for the others, or, equivalently, use a four level factor variable, with different levels for each of the last three days. With a deviance of 1.97 (3.97) and 3 d.f., our model now fits very well. Suicide is equally probable on the first four weekdays but varies among Friday, Saturday, and Sunday. We are, in fact, fitting the model exactly to these three observations, as can be seen from the residuals in Table 1.8. Thus, a more accurate value of the AIC would be 9.97, allowing three extra parameters for these days.

Through this series of simple frequency tables, we have now encountered many of the basic principles of model building for categorical data. We are

Table 1.8. Fitted values and residuals from several models for Durkheim's suicides in Table 1.7.

		Equiprobability		First weekday/weekend		Second weekday/weekend	
Day	Obs.	Fitted	Res.	Fitted	Res.	Fitted	Res.
Mon.	1001	941	1.956	1012.7	−0.369	1012.8	−0.369
Tue.	1035	941	3.064	1012.7	0.699	1012.8	0.699
Wed.	982	941	1.337	1012.7	−0.966	1012.8	−0.966
Thu.	1033	941	2.999	1012.7	0.636	1012.8	0.636
Fri.	905	941	−1.174	845.3	2.052	905.0	0.000
Sat.	737	941	−6.650	845.3	−3.726	737.0	0.000
Sun.	894	941	−1.532	845.3	1.674	894.0	0.000

ready to proceed to more complex models involving frequencies classified by several variables. However, we shall return to simple one-way classifications in Chapter 6.

1.6 Exercises

(1) Table 1.1 gave frequencies for subjects reporting only one stressful event. Below is the corresponding table (Haberman, 1978, p. 80) for all subjects, including those reporting several events, where one event has been selected at random.

Months Before	1	2	3	4	5	6	7	8	9
Number	49	55	42	43	35	35	42	31	37

Months Before	10	11	12	13	14	15	16	17	18
Number	21	35	40	29	22	29	12	15	15

(a) Fit an exponential decay model.
(b) Does this give a curve similar to that for the table in the text?

(2) The number of medical papers published each year from 1969 to 1985 in the statistical journal, *Biometrics*, have been tabulated by Sprent (1993, p. 61):

1969	1970	1971	1972	1973	1974	1975	1976	1977
11	6	14	13	18	14	11	22	19

1978	1979	1980	1981	1982	1983	1984	1985
19	25	24	38	19	25	31	19

Is there any evidence for an increasing or decreasing trend in the number of such publications in these data?

(3) Cases of acute lymphatic leukaemia were recorded, as reported to the British Cancer Registration Scheme, between 1946 and 1960 (Plackett, 1974, p. 8, from Lee):

January	40
February	34
March	30
April	44
May	39
June	58
July	51
August	55
September	36
October	48
November	33
December	38

Study the seasonal variation in these data.

(4) Haberman (1978, pp. 44, 51) also gives the suicides in the U.S.A. for 1969 and 1970. These are given in the following table, along with those from 1968 in Table 1.5.

Year	Jan.	Feb.	Mar.	Apr.	May	June
1968	1720	1712	1924	1882	1870	1680
1969	1831	1609	1973	1944	2003	1774
1970	1867	1789	1944	2094	2097	1981

	July	Aug.	Sept.	Oct.	Nov.	Dec.
1968	1868	1801	1756	1760	1666	1733
1969	1811	1873	1862	1897	1866	1921
1970	1887	2024	1928	2032	1978	1859

(a) Find a periodicity model for each of the 1969 and 1970 suicides.

(b) Do any of the models developed in Section 1.4 fit well to these two years of data? (A fixed known model can be fitted to new data by specifying it in an offset and fitting a null model.)

(5) Study the symmetry of political views of subjects (Haberman, 1978, p. 85), as expressed in the following table:

Extremely liberal	46
Liberal	179
Slightly liberal	196
Moderate, middle of the road	559
Slightly conservative	232
Conservative	150
Extremely conservative	35

(6) Jarrett (1979) tabulated the intervals between explosions in British mines between 1851 and 1962. He summarized the distribution of the number of accidents each day of the week as follows:

Sunday	5
Monday	19
Tuesday	34
Wednesday	33
Thursday	36
Friday	35
Saturday	29

Study how such accidents vary over the week.

(7) The numbers of passengers boarding the 11:13 train from Fredericia, Denmark, to Copenhagen were recorded in February 1976 (Andersen, 1980, p. 136):

Sunday	429
Monday	288
Tuesday	270
Wednesday	262
Thursday	279
Friday	126
Saturday	130

Study how the number of passengers varies over the week.

(8) Jarrett (1979) also summarized the mine explosions, mentioned in Exercise (1.6), by month:

January	14
February	20
March	20
April	13
May	14
June	10
July	18
August	15
September	11
October	16
November	16
December	24

Is there any systematic change over the year?

(9) The distribution of traffic accidents involving pedestrians in Denmark for 1981 (Andersen, 1991, p. 59) was recorded as follows:

Sunday	130
Monday	279
Tuesday	256
Wednesday	230
Thursday	304
Friday	330
Saturday	210

Develop a simple model to describe differences over the week.

(10) The table below gives suicides by region of the U.S.A. in 1970 (Haberman, 1978, p. 53).

Region	Suicides	Pop. (100 000)
New England	1119	118.41
Middle Atlantic	3165	371.99
East North Central	4308	402.52
West North Central	1790	163.19
South Atlantic	3729	306.71
East South Central	1335	128.03
West South Central	2075	193.21
Mountain	1324	82.82
Pacific	4635	265.23

Determine if the suicide rate is the same in all regions.

2
Larger tables

In Chapter 1, we studied some models for patterns in one-way tables. These methods can often be extended directly to larger tables. Here, we shall be interested in the dependence of a response distribution on explanatory variables in a variety of contexts. Thus, as often in statistics, we shall distinguish between response variables and explanatory variables and look for patterns in the dependence of (the distribution of) the former on the latter. In some cases, we may even be able to expect a certain precise pattern of dependence for theoretical reasons; in the latter context, we shall conclude the chapter by looking at an example from genetics.

It is useful, at this stage, to consider the different ways in which studies may be conducted. We may distinguish

- prospective studies, where explanatory variables are observed and the subjects followed up to record values of the response;
- retrospective studies, where subjects are selected according to values of the response variable, then values of the appropriate explanatory variables are traced back in time;
- cross-sectional studies, where all variables are recorded more or less simultaneously.

Log linear and logistic models are appropriate for all three types of study. In fact, for retrospective studies, they are the only types of model to provide satisfactory results.

2.1 Dependence

With more than one variable to classify events, the table of frequencies will have several dimensions. A first question is whether there is dependence of the response on the other variables. In other words, we are going to look at relationships among different classifications of events, the variables. Note that the events enumerated within a table can usually be assumed to be independent if they occur for different individuals who are not influencing each others' responses.

We shall look at dependence, in a simple example with two variables, in this section. If dependence is discovered, the next step is to try to find some structure in it. That will be the subject of the rest of the book.

Table 2.1. Political affiliation of college students in the U.S.A. (Christensen, 1990, p. 52)

	Political affiliation		
College	Republican	Democrat	Independent
Literature	34	61	16
Engineering	31	19	17
Agriculture	19	23	16
Education	23	39	12

With software for generalized linear models, the Poisson representation of the multinomial distribution, presented in Section 1.1, must be used. This means that a cross-tabulated table, to which log linear models are to be fitted, must be stored as a single vector containing the observed frequencies. A series of other vectors will then be defined to index the rows, columns, and so on, of the multi-dimensional matrix. Each of these latter vectors represents a variable to be fitted to the data. Thus, analysis of log linear models with such software involves one more vector of values than the number of variables to be included in the model. For example, a two-way table has two variables, but will require three vectors.

Consider a simple two-way table relating the response, political affiliation (`AFFIL`), to the explanatory variable, college enrollment (`ENROL`) of university students in the U.S.A., as presented in Table 2.1. We first fit a model where only the two sets of marginal totals, i.e. the row and column sums, are fixed,

$$\log(\nu_{ij}) = \mu + \theta_i + \phi_j$$

where i indexes political affiliation and j college enrollment. (When there is a response variable in a table, it will be indexed by i and explanatory variables will be indexed by following letters in the alphabet.) Then, this model has these two sets of marginal totals as sufficient statistics. This is the *minimal* model for these data. It represents independence between affiliation and college, and may thus be used to verify whether there is a relationship between these two variables. In the terminology used by generalized linear modelling software, this can be fitted as

$$\texttt{AFFIL} + \texttt{ENROL} \qquad (2.1)$$

Each of the (mean) parameter vectors is analogous to those already encountered in Chapter 1. With a deviance of 16.39 (28.39) and 6 d.f., this independence model is implausible.

If we inspect the plot of Cook's distances in Figure 2.1, we see that the model fits poorly to observations 2 and 5, corresponding to students

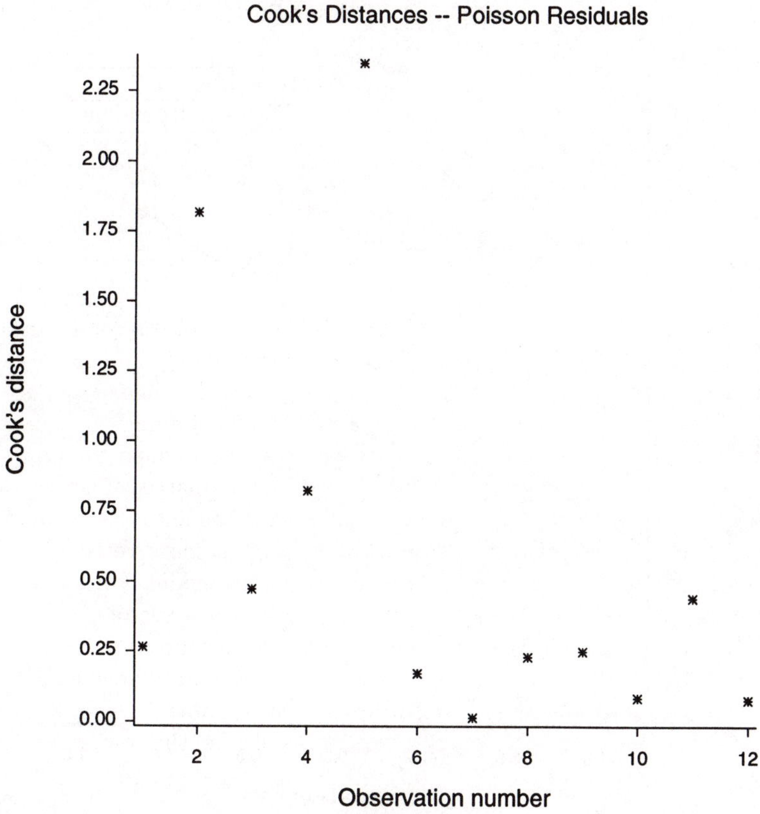

Fig. 2.1. Cook's distances from the independence model of Equation (2.1) for the political affiliation data of Table 2.1.

of literature and engineering with a Democrat affiliation. (In all multi-way tables, observations will be numbered across the rows, in the order in which a computer reads them.) Also, the normal probability plot, in Figure 2.2, is not very straight, lying at greater than 45 degrees. Thus, both plots confirm our conclusion that the model fits poorly.

An alternative is to fit what is known as the saturated model, a model with as many parameters as there are entries in the table:

$$\log(\nu_{ij}) = \mu + \theta_i + \phi_j + \gamma_{ij} \tag{2.2}$$

This model must necessarily fit the data exactly; the deviance will be zero (AIC 24). Here γ_{ij} refers to a matrix of parameters describing the full dependence relationship between political affiliation and college enrollment. In the same way as for the mean parameter vectors, constraints must be

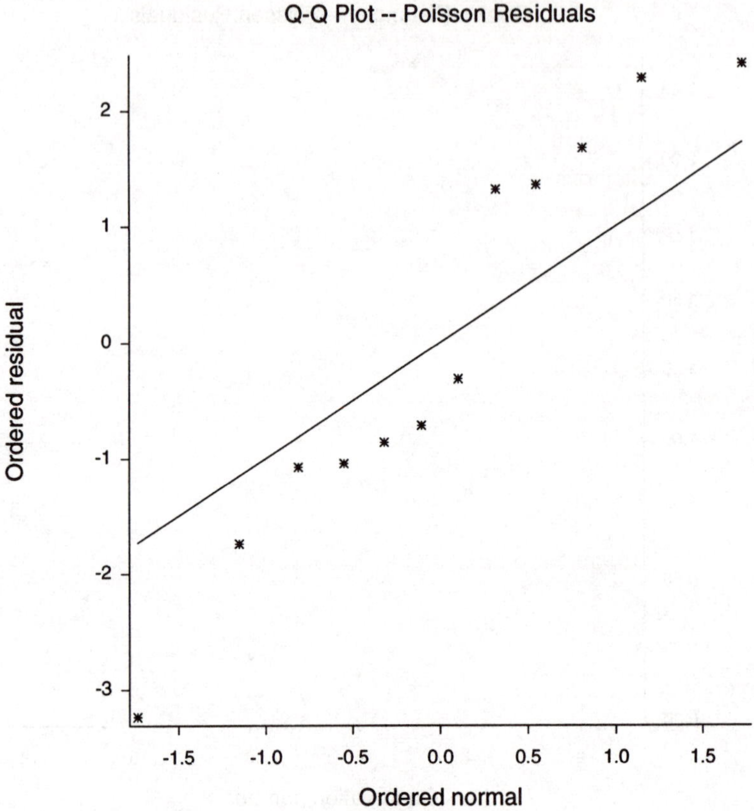

Fig. 2.2. Normal probability plot from the independence model of Equation (2.1) for the political affiliation data of Table 2.1.

applied to this parameter matrix in order to be able to estimate the model. As with the factor variables in Chapter 1, generalized linear model software may simply set one row and one column of γ_{ij}, often the first, to zero so that all remaining values are comparisons with that category of each variable. These are called baseline constraints. Here, the complications of interpretation begin. But the model may very easily be fitted by placing a dot operator between the two variable names

$$\texttt{AFFIL} + \texttt{ENROL} + \texttt{AFFIL} \cdot \texttt{ENROL}$$

The three terms correspond exactly to the last three in the log linear model of Equation (2.2) above. Most such software fits the constant term implicitly, so that it need not be shown explicitly in the formula. Another

Table 2.2. Parameter estimates from the saturated model of Equation (2.3) for the political affiliation of Table 2.1.

	Estimate	s.e.
AFFIL		
Republican	0.000	—
Democrat	0.585	0.214
Independent	−0.754	0.303
ENROL		
Literature	0.000	—
Engineering	−0.092	0.248
Agriculture	−0.582	0.286
Education	−0.391	0.270
AFFIL·ENROL		
Democrat·Literature	0.000	—
Democrat·Engineering	−1.074	0.362
Democrat·Agriculture	−0.394	0.377
Democrat·Education	−0.056	0.339
Independent·Literature	0.000	—
Independent·Engineering	0.153	0.428
Independent·Agriculture	0.582	0.455
Independent·Education	0.103	0.469

equivalent formulation is

$$\texttt{AFFIL} * \texttt{ENROL} \tag{2.3}$$

The use of the asterisk operator implies that all lower order terms are automatically included, called a hierarchical model. Note that this is not the same as multiplying AFFIL*ENROL in a calculation and then fitting the result. The latter would add only one (linear interaction) term to the model, while the former adds a (large) number of main effect and interaction terms.

The parameter estimates for this saturated model are given in Table 2.2. Here, as in tables throughout the book, all contrasts are with respect to the first category of each variable, which has been taken to be the baseline. Independent affiliation, as compared to Republican, occurs more often among engineering, agriculture, and education students than among students of literature, as seen by the positive interaction values, while Democrats are less often there, as shown by the negative interaction values. However, the standard errors indicate that only the Democrat·engineering interaction may be worth retaining. This is supported by what we observed with the residuals above.

Although this baseline constraint approach is a standard type of output

Table 2.3. Mode of travel, frequency of trip, and size of centre visited in the English Midlands. (Fingleton, 1984, p. 85)

Centre	Mode	Frequent	Seldom
Small	Walk	102	28
	Bus	2	3
	Car	9	10
Medium	Walk	48	12
	Bus	10	11
	Car	3	15
Large	Walk	43	18
	Bus	27	45
	Car	22	89

from generalized linear modelling software for dependence in cross-classified tables, we see that it can be rather difficult to interpret. Another way of studying the same dependence relationships, the conventional constraints, will often be used below.

For these data, little more structure can be placed on the dependence because the variables are nominal. One possibility is to try to collapse a table to a smaller size, so that categories with the same profile of probabilities of response, or responses always having the same probability, are each grouped together, something we look at in the next section.

2.2 Collapsing tables

If there are several variables in a model, one way to simplify it is to try to remove some of them. This can be done, for example, if the response is completely independent of an explanatory variable. It has the effect of collapsing the table to one of smaller dimension. A second possibility, when variables have more than two categories, is to try to amalgamate the categories of a variable. This reduces the size of the table, but without changing its dimension because no variables are eliminated.

Let us look at a simple example, a three-way table concerning the binary response, frequency of `TRIPS` made, as it depends on the explanatory variables, `MODE` of travel and `SIZE` of the centre visited, given in Table 2.3. The model for independence for such data can be represented as the log linear model

$$\texttt{TRIPS} + \texttt{MODE} * \texttt{SIZE}$$

Because `TRIPS` does not interact with the two explanatory variables in this model, it is independent of them. This model has a deviance of 132.90 (134.90) with 8 d.f., indicating that it is unacceptable.

Table 2.4. Parameter estimates from the model of Equation (2.5) for dependence of 'seldom' `TRIPS` on `MODE` and `SIZE` for the mode of travel data of Table 2.3.

	Estimate	s.e.
TRIPS·MODE		
Seldom·Walk	0.000	—
Seldom·Bus	1.352	0.281
Seldom·Car	2.167	0.266
TRIPS·SIZE		
Seldom·Small	0.000	—
Seldom·Medium	0.193	0.307
Seldom·Large	0.583	0.273

The addition of dependence on `SIZE`

$$\mathtt{TRIPS} + \mathtt{MODE} * \mathtt{SIZE} + \mathtt{TRIPS} \cdot \mathtt{SIZE}$$

reduces this to 80.06 (86.06) with 6 d.f. If, instead, dependence on `MODE` is added,

$$\mathtt{TRIPS} + \mathtt{MODE} * \mathtt{SIZE} + \mathtt{TRIPS} \cdot \mathtt{MODE} \qquad (2.4)$$

the deviance is 9.02 (15.02) with 6 d.f., an acceptable model. Reintroducing dependence on `SIZE`

$$\mathtt{TRIPS} + \mathtt{MODE} * \mathtt{SIZE} + \mathtt{TRIPS} \cdot (\mathtt{MODE} + \mathtt{SIZE}) \qquad (2.5)$$

gives a model with a deviance of 4.14 (14.14) with 4 d.f. The change in deviance confirms this conclusion, but the AIC indicates that the latter model is preferable. The parameter estimates for dependence of `TRIPS` on `MODE` and `SIZE` are given in Table 2.4. Suppose that we conclude that, when frequency of travel is the response of interest, the table may be collapsed and the variable describing the size of centre visited removed.

The residual plots for the model without the size of centre may now be studied. Cook's distances, in Figure 2.3, show that the two frequencies for observations 12 and 18, seldom going by car to a medium or large centre, have values much greater than one and are fitted poorly by the model. The normal probability plot, in Figure 2.4, has two extreme values, one at each end. The negative one is observation 12, but the positive one is observation 3, travelling frequently by bus to a small centre. Thus, it is also interesting to look at the corresponding plots for the model with size of center. Cook's distances, in Figure 2.5, now indicate that the first frequency, going frequently on foot to a small centre, is fitted poorly by the model. The normal probability plot, in Figure 2.6, shows no abnormalities.

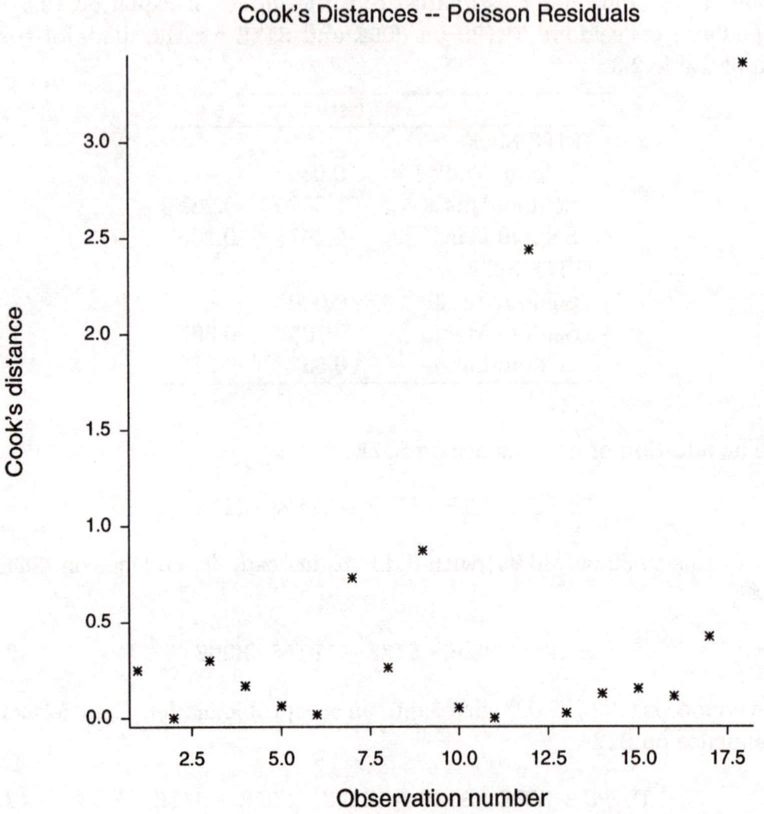

Fig. 2.3. Cook's distances for the mode of travel study of Table 2.3 from the model of Equation (2.4) without centre size.

We may then conclude that further inspection of the variable, size of centre, which we plan to remove, is worthwhile. In the models including it, the parameter estimate relating frequency to medium centres is of similar size to that for small centres, as seen in Table 2.4. Thus, we can try a model combining small and medium centres. When we refit the model of Equation (2.5) with only two categories for `SIZE` in this way, we find that the deviance is now 4.54 (12.54) with 5 d.f. Thus, eliminating the two category `SIZE` variable from this model, i.e. going back to the model of Equation (2.4), would result in the deviance increasing by (9.02 – 4.54 =) 4.48 with 1 d.f.

Our final conclusion is that we should not collapse the table to two dimensions by completely eliminating the centre size variable. Instead the size of the table can be reduced by amalgamating the first two categories

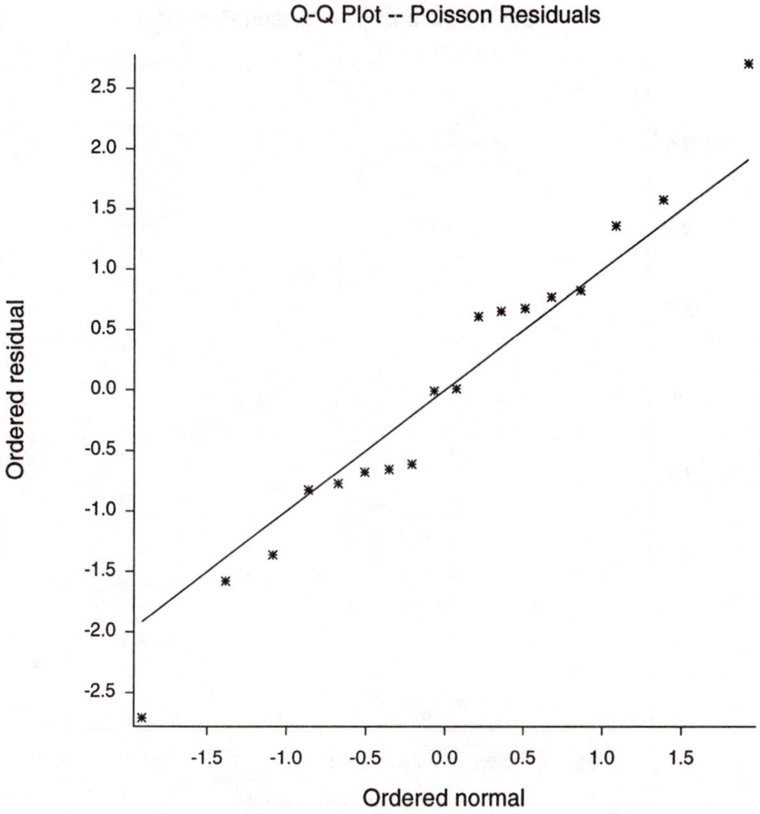

Fig. 2.4. Normal probability plot for the mode of travel study of Table 2.3 from the model of Equation (2.4) without centre size.

of this variable. Frequency of trips depends on mode of travel (0.000, 1.392, 2.188), with 'seldom' visits more often by car, and also on the size of the centre (0.000, 0.489), with more people having 'seldom' visits to large centres.

In general, the possibility of reducing the size of a table can readily be determined by standard procedures of variable checking and elimination. This form of model simplification will be especially important as the number of variables in a table grows.

2.3 Prospective studies

An eternal problem in the human sciences is that of determining direction of causality. Although perhaps most discussed in the social sciences, it is not unique to those disciplines, because it also has the same importance, for

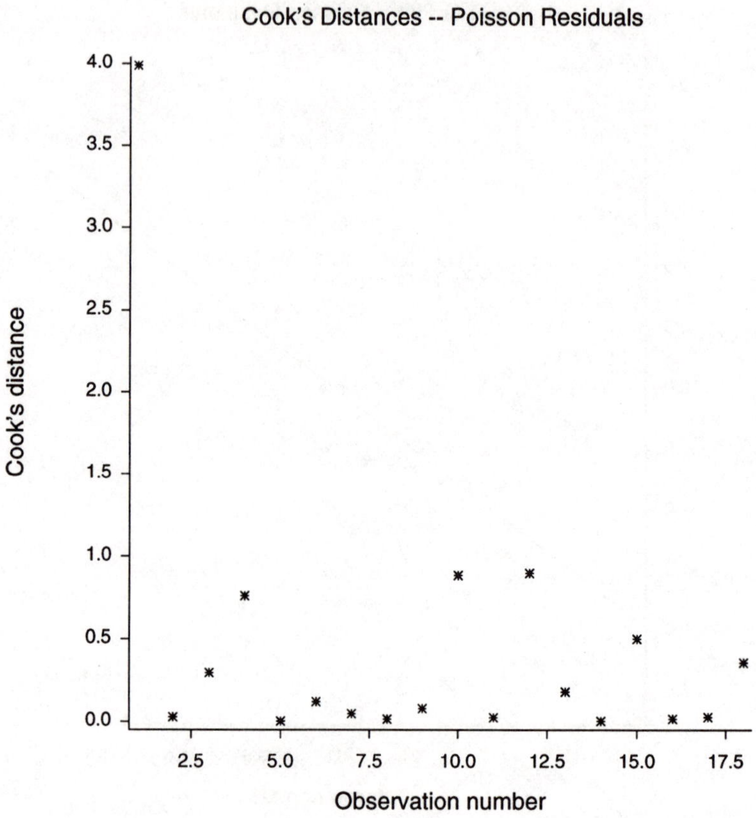

Fig. 2.5. Cook's distances for the mode of travel study of Table 2.3 from the model of Equation (2.5) with centre size.

example, in the medical sciences, and even in some natural sciences, such as in astronomy. Most of the natural sciences are able to resolve the problem through the application of experimental methods. Such is not often possible for the human sciences. Thus, usually, no unequivocal means is available to determine direction of causality where experimentation must be excluded. Various indirect methods must be applied. The more mutually-confirming approaches used, the more confident may we become that we are perhaps succeeding in isolating a cause.

One of the most useful approaches to the problem of studying causality in this context is the use of a time factor. Events which occur later in time cannot affect earlier events, or at least we may so suppose in most cases. As we have seen, two approaches to collecting chronological information may be distinguished: (1) we may choose a sample of individuals accord-

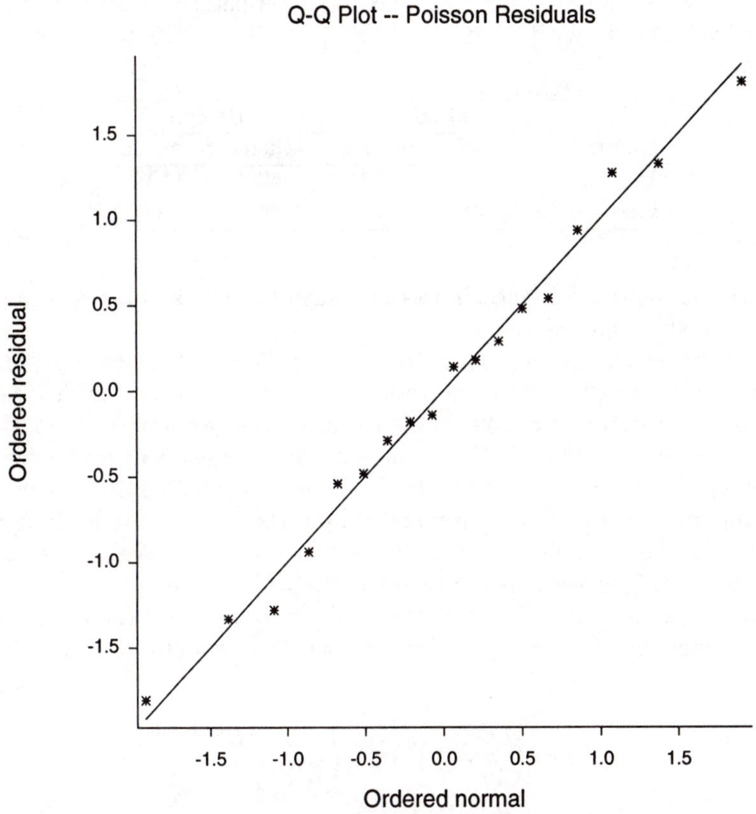

Fig. 2.6. Normal probability plot for the mode of travel study of Table 2.3 from the model of Equation (2.5) with centre size.

ing to the criteria of certain explanatory variables and then follow them up in time to see what response variable, the variable to be explained, is obtained, or (2) we may choose a sample according to the response variable and then investigate what values of the explanatory variables had previously (in time) existed. The first case is a prospective study. We shall consider it in this section. It also includes panel studies, where the same response variable is observed several times, which we shall look at in Part II. The second is a retrospective study, which we shall look at in the next section; in the medical sciences, this is often called a case-control study. We may note that the first approach resembles experimentation in the natural sciences, with, however, absence of random allocation of the explanatory variables. And, in fact, the methods of statistical analysis are often identical, although the strength of conclusions cannot be. In contrast,

Table 2.5. Three-year survival of breast cancer patients according to nuclear grade and diagnostic centre. (Whittaker, 1990, p. 220, from Morrison *et al.*)

	Malignant		Benign	
Centre	Died	Survived	Died	Survived
Boston	35	59	47	112
Glamorgan	42	77	26	76

the second approach is specific to the human sciences and often requires special analytic procedures.

One important use of prospective studies is in the follow-up of patients in medical research. Often, the response variable is binary, as in recovered/ill or died/survived. We shall look at a simple example of the latter case, as given in Table 2.5. Cancer patients in two treatment centres were followed up over a three-year period to determine who survived. One of the important factors distinguishing them was their nuclear grade. Interest centres on the differential probability of survival for the two grades, and on whether this differs between the two diagnostic centres.

When the response is binary, the easiest way to model the data is to use the binomial distribution. With the logit link (π and n as in Chapter 1),

$$\log\left(\frac{\pi_1}{1-\pi_1}\right) = \log\left(\frac{\nu_1}{n-\nu_1}\right)$$

this logistic model will give the same results as a log linear model, as we shall see in the next section. A logit is the log of the odds ratio of the probability of one response with respect to the other. This model has the advantage of being substantially simpler to fit; as we shall see, vectors are shorter and fewer interactions must be included. For this approach, the frequencies must be contained in two vectors (each being one half the length that they would be with a log linear model). These may correspond to the two possible responses, here, died and survived. However, in specifying the binomial distribution for generalized linear modelling software, we require a different vector, the total number concerned under each combination of the explanatory variables, called the binomial denominator. Thus, the second vector can correspond to either of the response categories; the choice will only reverse all of the signs of the parameters. (The reader may try what follows both ways to verify this.) Here, we shall specify `DIED` as the 'response' vector and `TOTAL` as the binomial denominator with a binomial distribution.

Let us begin by fitting the fullest model possible, the saturated model:

Table 2.6. Parameter estimates and standard errors from the saturated model of Equation (2.6) for the cancer survival data of Table 2.5.

	Estimate	s.e.
1	−0.522	0.213
CENTRE	−0.084	0.287
GRADE	−0.346	0.275
CENTRE·GRADE	−0.120	0.405

$$\texttt{CENTRE} * \texttt{GRADE} \tag{2.6}$$

As with all saturated models, this has a zero deviance (AIC 8). The parameter estimates, with the first category of each variable as baseline, are given in Table 2.6.

In complex models, we may start, in a hierarchical fashion, by looking at the highest order interaction terms to see if they can be eliminated. There are at least two good reasons for this. Such interactions are complicated and difficult to interpret. Also, they usually have little meaning if less complex terms, which they contain, are first eliminated from the model. Thus, if, say, CENTRE were eliminated, the interaction, CENTRE·GRADE, would no longer have much meaning. Conversely, if the interaction is required in the model, elimination of a lower order term yields a model which is usually not interpretable. Such a series of models is called hierarchical. If we look at Table 2.6, we see that the interaction should not be necessary in the model, because $|-0.120/0.405| < 2$. When we remove it to obtain the model,

$$\texttt{CENTRE} + \texttt{GRADE}$$

the deviance only rises by 0.09 (AIC 6.09), confirming our suspicion. This means that, if both variables remain in the model, survival depends on grade in the same way in both centres.

From the parameter estimates and standard errors of this new model (not shown), we suspect that CENTRE can also be removed. When this is done, the deviance only changes by a further 0.51 (AIC 4.60). We are left with a model where survival only depends on the nuclear grade. The parameter estimate is −0.337 (s.e. 0.198), indicating that the odds, or probability ratio, of dying to survival is less in the second nuclear grade, benign, than in the first, malignant. Notice how much smaller the standard error now is than it was in Table 2.6, although the parameter estimate has not changed very much. This generally happens as a model is simplified by removing unnecessary components. A final check, removing the effect of GRADE, increases the deviance by 3.62 (AIC 6.22), indicating that this variable should be left in the model (and that the standard error is misleading).

Table 2.7. Clinic use, attitude to extra-marital sex, and virginity. (Fienberg, 1977, p. 92, from Reiss *et al.*)

		Use clinic	
Virgin	Attitude to sex	Yes	No
Yes	Always wrong	23	23
Yes	Not always wrong	29	67
No	Always wrong	127	18
No	Not always wrong	112	15

With only four observations, residual plots are of relatively little use here.

2.4 Retrospective studies

Our example of a retrospective study is a case-control study, as given in Table 2.7. The response variable of interest is `USE` of a university contraceptive clinic. A sample of clinic users formed the case group and a corresponding sample of non-users formed the control. The data were checked to verify that the two groups were similar for various background variables. Often in case-control studies, the individuals are actually matched pairs on these variables, in which case the model should take this into account (see Sections 10.3 and 10.4).

The explanatory variables of interest are virginity (`VIRGIN`) and attitude to extra-marital sexual relations (`ATTITUDE`). Each of the three variables involved is binary or dichotomous. As we saw in the previous section, when the response variable is binary, analysis may be simplified by using a binomial distribution instead of Poisson. For this example, we shall present the two approaches to demonstrate that the results are the same and to illustrate the relationships between them. We begin with the analysis using the Poisson distribution, that is, with the log linear model:

$$\log(\nu_{ijk}) = \mu + \theta_i + \phi_j + \omega_k + \alpha_{ij} + \beta_{ik} + \psi_{jk} + \gamma_{ijk} \qquad (2.7)$$

where i indexes clinic use, j virginity, and k attitude, or

`USE * VIRGIN * ATTITUDE`

We shall consider here only the conventional constraints of summation to zero. Thus, for dichotomous variables, the design matrix can be constructed very simply, by assigning values of 1 and -1 to the two categories of each variable. We now have a much more complex log linear model than previously because, with more than two variables present, they can interact in groups of two or more, as we can see in Equation (2.7).

With a log linear model for frequency data, we generally consider that the marginal totals for the three variables are fixed, ensured by the terms in

Equation (2.7) with single indices, as are the totals for the relationship between the two explanatory variables, ATTITUDE and VIRGIN (ψ_{jk}). Again, these totals are sufficient statistics for the parameters in the model. As for one-way tables, this guarantees that the probabilities sum to unity for the response distribution under each possible combination of conditions. Instead of starting with the saturated model, as in the previous section, here we shall build up from the simplest model. Thus, our minimal model, expressing independence between clinic use and the two explanatory variables, is

$$\log(\nu_{ijk}) = \mu + \theta_i + \phi_j + \omega_k + \psi_{jk}$$

which may be written

$$\texttt{USE} + \texttt{VIRGIN} * \texttt{ATTITUDE}$$

After fitting, we see that this model, with a deviance of 121.34 (131.34) and 3 d.f., must be rejected.

Clinic use depends either upon attitude or upon virginity or both. We now introduce the relationship between clinic use and virginity:

$$\texttt{USE} + \texttt{VIRGIN} * \texttt{ATTITUDE} + \texttt{USE} \cdot \texttt{VIRGIN}$$

and immediately find that we have an acceptable model. The deviance is 5.19 (17.19) with 2 d.f. Virgins have a lower probability of using the clinic, as indicated by the negative value of the parameter for the USE·VIRGIN interaction, −0.632.

If, instead of this relationship, we substitute that between clinic use and attitude,

$$\texttt{USE} + \texttt{VIRGIN} * \texttt{ATTITUDE} + \texttt{USE} \cdot \texttt{ATTITUDE}$$

we see that the model is not acceptable. It has a deviance of 109.60 (121.60), again with 2 d.f. Finally, if we include both relationships at the same time,

$$\texttt{USE} + \texttt{VIRGIN} * \texttt{ATTITUDE} + \texttt{USE} \cdot \texttt{ATTITUDE} + \texttt{USE} \cdot \texttt{VIRGIN}$$

the deviance is 2.92 (16.92) with 1 d.f. We again have an acceptable model, slightly better than that with only the relationship between clinic use and virginity. If we prefer a simple model, we might exclude the relationship between clinic use and attitudes to sex outside marriage. Clinic use would then depend only on virginity in these data.

We can now repeat the analysis using the binomial distribution. Our model is now

$$\log\left(\frac{\nu_{1jk}}{\nu_{2jk}}\right) = \log\left(\frac{\pi_{1jk}}{\pi_{2jk}}\right)$$
$$= \omega + \alpha_j + \beta_k + \gamma_{jk} \qquad (2.8)$$

or

`VIRGIN * ATTITUDE`

where the four parameter sets here correspond to those with the same Greek symbols in the model of Equation (2.7) above. We again use the conventional constraints. With the model presented in this way, in its logistic form, we may interpret the relationship between the three variables at the same time (γ), as the statistical interaction between the two explanatory variables with respect to the response variable.

As we saw in the previous section, in the case of the binomial distribution, specification of the model distribution requires an additional vector, that of the binomial denominator. The data must be presented somewhat differently from those for the log linear model above, because we now have two vectors containing the observed frequencies, that is, users and the total for each combination of categories of the explanatory variables.

If we follow the same steps as above, we fit the series of models

- `1`
- `VIRGIN`
- `ATTITUDE`
- `VIRGIN+ATTITUDE`

We shall see that all of the deviances are the same as for the log linear model but that the corresponding parameter values of interest are twice as large for the logistic model of Equation (2.8) as for the log linear model of Equation (2.7). (With other constraints, they may be identical.) Model fitting is simpler and more efficient because the minimal model is the null model, so that there are always fewer variables in the fit, and the vectors are half as long.

We have analysed this example as if the relationships between response and explanatory variables were clear. A moment's reflection, however, shows that no specific time sequence necessarily holds among the variables. Any one of the three might be the response, explained by the other two. Indeed, our first analysis, with the log linear model, permits any one of the three interpretations, although the analysis would proceed slightly differently due to choice of the minimal model and the subsequent inclusion of the relationship between a different pair of variables in it. Or all three variables may be considered on the same level, as a multivariate response distribution (Section 2.6 below), and their inter-relationships studied. In that case, only the three main effects would be included in the minimal model. Again, the log linear model permits this. Response and explanatory variables are distinguished only by the interpretation of the analyst and not by the statistics of the analysis. This is directly related to the fact that the log linear (and logistic) model may be equally well applied to prospective and retrospective studies.

Our final model must, at least, include the relationships between attitude and virginity (the reader is invited to verify that this is necessary) and between virginity and clinic use. Six or more distinct lines of causality may be imagined. These may be represented by interaction graphs:

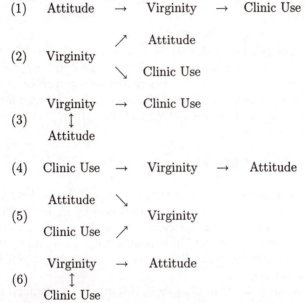

(1) Attitude → Virginity → Clinic Use

(2) Virginity ↗ Attitude
Virginity ↘ Clinic Use

(3) Virginity → Clinic Use
Virginity ↕ Attitude

(4) Clinic Use → Virginity → Attitude

(5) Attitude ↘ Virginity
Clinic Use ↗ Virginity

(6) Virginity → Attitude
Virginity ↕ Clinic Use

Here, a single-headed arrow indicates causality and a double-headed arrow indicates a non-causal relationship of association. Such path diagrams are often useful in summarizing certain simple relationships. However, their application is limited, because, for the representation to be clear, all variables should be dichotomous and no interactions may be present. Such a situation is rarely the case in the study of any complex social or biological phenomena.

Such path diagrams should not be confused with graphical models (Whittaker, 1990) which can represent (at least some) interactions, but not certain other simpler hierarchical models. Thus, with three variables, say `A`, `B`, and `C`, an interaction graph with lines linking all three would represent `A*B+A*C+B*C` if it were a path diagram and `A*B*C` if it were a graphical model. The latter are most useful for strictly multivariate data, where all variables are responses, with no explanatory variables present.

In addition, the user must be wary of interpreting such diagrams as demonstrating causality. In this series of causal models, we have not even considered the possibility that some other factor, not included in the study, such as, for example, personality, social class, or biological makeup, is, in fact, the underlying causal factor for two or all three of these variables. In no case, whether in making a choice among the six or more alternative causal models or in considering an external factor, can the analysis with the log linear model aid in making the decision. The choice must depend on other outside information, perhaps combined with knowledge of the way the data were collected. *Post hoc* statistical analysis cannot resolve problems of causality.

Formulae for multi-dimensional tables

We may now generalize the fitting procedure to a table of any dimension. Suppose that the response variables are called R1, R2, ... and the explanatory variables E1, E2, ... (there may be one or more response and zero or more explanatory variables). Then the general formula for independence of all responses from each other and from the explanatory variables in log linear models is

$$\mathtt{R1} + \mathtt{R2} + \cdots + \mathtt{E1} * \mathtt{E2} * \cdots \tag{2.9}$$

Such a minimal model guarantees that the response variables have proper probability distributions under all observed conditions, defined by the explanatory variables. This is a conditional model, so that, when several response variables are present, in a multivariate distribution, the conditional, not the marginal, distribution of each response, given the other responses, is being studied.

In fact, as we shall see in Chapter 6, this is only the minimal model if a specific probability distribution function is not being fitted to the response. Thus, a response variable, say R1, need not be a factor variable; if the response is quantitative, it may be some function(s) of that quantity. If a response is omitted completely, we have a uniform distribution for that variable, the equiprobability model of Chapter 1.

In the log linear approach, dependencies are obtained by adding the product of the desired variables, as in R2·E3. When there is only one response variable and it is binary, a logistic model can be used, with the same results. Then, as we have seen, independence is given by the null model and dependencies are introduced by adding the appropriate variables.

In this section, we started from the minimal model and searched for variables which might be added to it. This is often a reasonable scientific procedure, because we usually hope for a simple model to explain the observed phenomena. However, arguments can also be made for starting from a complex model and eliminating terms, as we did in the previous section. Unfortunately, the two approaches may not lead to the same final model because changes in deviance for adding or eliminating a given term will depend on what other terms are present. This contrasts with the analogous normal theory ANOVA for balanced designs, where terms are often orthogonal and order of elimination makes no difference.

2.5 Cross-sectional studies

Often in a study, all variables are observed at the same moment in time, that is, a cross-sectional study (although many are, in fact, retrospective). Many such variables may be recorded, and it is often necessary to sort through them to ascertain which have explanatory power. One reasonable procedure is to start with all main effects, simultaneously checking if they are all necessary and if interactions need be added.

Let us look at results from a cross-sectional health survey in one large cotton textile company. Six variables from this study are cross-classified in Table 2.8. Notice the considerable number of zeros in the table. We are interested in determining under what conditions workers suffer more from byssinosis. With a binary response, we can use a logistic model.

When we fit this logistic model with only main effects,

$$\texttt{DUST} + \texttt{RACE} + \texttt{SEX} + \texttt{SMOKE} + \texttt{EMPLOY} \qquad (2.10)$$

we obtain a deviance of 43.27 (59.27) with 57 d.f., a very acceptable fit. (Note that seven combinations of the explanatory variables have no observations, hence the reduction in degrees of freedom.) We apparently do not require any interaction parameters in the model. The parameter estimates are given in Table 2.9. Inspection of the standard errors indicates that both `SEX` and `RACE` may be eliminated, yielding the model,

$$\texttt{DUST} + \texttt{SMOKE} + \texttt{EMPLOY} \qquad (2.11)$$

This increases the deviance by only 0.61 (AIC 55.88). On the other hand, eliminating `DUST`, `SMOKE`, or `EMPLOY` raises it, respectively, by a further 247.09 (AIC 298.97), 11.43 (AIC 65.32), or 14.53 (AIC 66.41). The parameter estimates in this model change very little from those given in Table 2.9. However, the medium and low dust levels have very similar parameter values. Let us combine them by giving both the factor level two. The change in deviance is only 0.34 (AIC 54.22). In the same way, the two longer employment periods are similar, so that we may also be able to regroup them. The resulting increase in deviance is 0.49. This final model, which has the same form as Equation (2.11) but where `DUST` and `EMPLOY` have only two categories each, has a deviance of 44.71 (52.71) with 61 d.f.

Binomial residuals are differences between observed and fitted values divided by their estimated standard errors, $\sqrt{n\hat{\pi}_i(1-\hat{\pi}_i)}$. (In the plots, they are further corrected by the appropriate element of the hat matrix.)

Table 2.8. Number of cotton workers suffering from byssinosis, as it depends on race, sex, smoking, length of employment (years) in the cotton industry, and dust level. (Higgins and Koch, 1977)

				Dust					
				High		Medium		Low	
				Suffer from byssinosis					
Race	Sex	Smokes	Empl.	Yes	No	Yes	No	Yes	No
White	Male	Yes	< 10	3	37	0	74	2	258
Other				25	139	0	88	3	242
White	Female			0	5	1	93	3	180
Other				2	22	2	145	3	260
White	Male	No		0	16	0	35	0	134
Other				6	75	1	47	1	122
White	Female			0	4	1	54	2	169
Other				1	24	3	142	4	301
White	Male	Yes	10–20	8	21	1	50	1	187
Other				8	30	0	5	0	33
White	Female			0	0	1	33	2	94
Other				0	0	0	4	0	3
White	Male	No		2	8	1	16	0	58
Other				1	9	0	0	0	7
White	Female			0	0	0	30	1	90
Other				0	0	0	4	0	4
White	Male	Yes	> 20	31	77	1	141	12	495
Other				10	31	0	1	0	45
White	Female			0	1	3	91	3	176
Other				0	1	0	0	0	2
White	Male	No		5	47	0	39	3	182
Other				3	15	0	1	0	23
White	Female			0	2	3	187	2	340
Other				0	0	0	2	0	3

The normal probability plot in Figure 2.7 shows some curvature in the lower tail, which might be investigated. It seems to indicate that the model fits too well in conditions with low probability of byssinosis. The plot of Cook's distances, in Figure 2.8, shows one value to be larger than the others, high dust for white male smokers, but a value of 0.45 is not excessive. Thus, these diagnostics confirm that the model is satisfactory.

In this simplified model, the remaining parameter estimates are somewhat changed from those in Table 2.9. The dust level, with parameter estimate -2.668 (s.e. 0.170) for medium and low levels, as compared to

Table 2.9. Parameter estimates from the model of Equation (2.10) for the health study of Table 2.8.

	Estimate	s.e.
1	−1.945	0.233
DUST(2)	−2.580	0.292
DUST(3)	−2.731	0.215
RACE(2)	0.116	0.207
SEX(2)	0.124	0.229
SMOKE(2)	−0.641	0.194
EMPLOY(2)	0.564	0.262
EMPLOY(3)	0.753	0.216

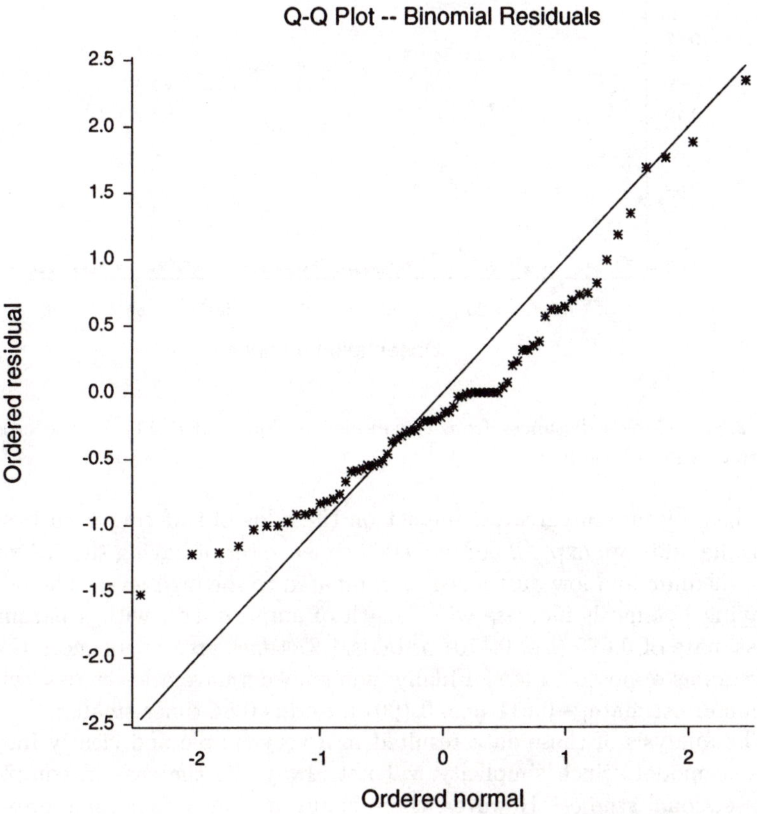

Fig. 2.7. Normal probability plot from the model of Equation (2.11) for the cross-sectional study of Table 2.8.

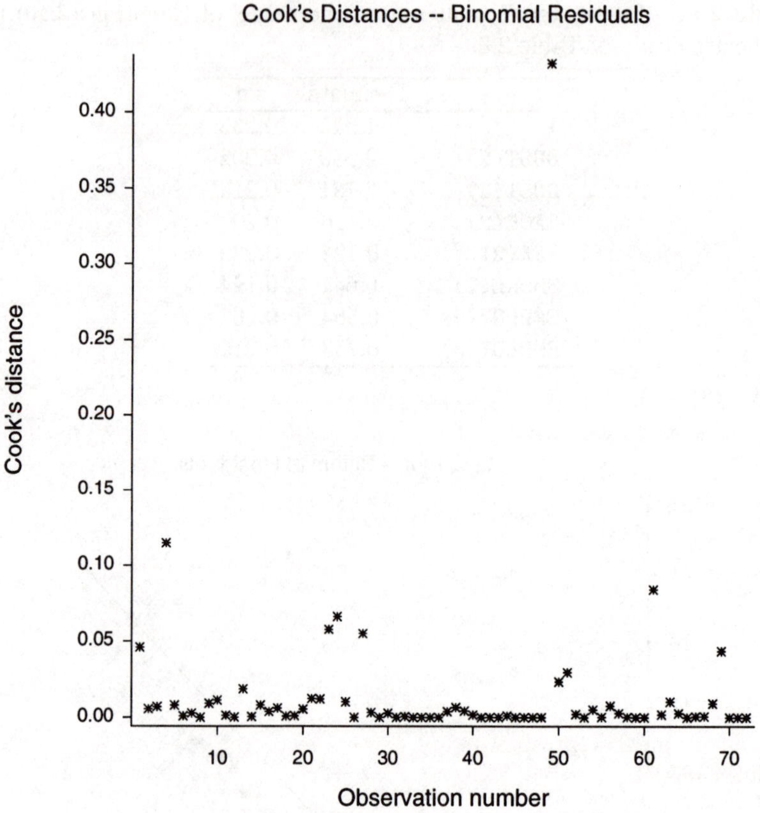

Fig. 2.8. Cook's distances from the model of Equation (2.11) for the cross-sectional study of Table 2.8.

high, has, by far, the greatest impact on the odds of suffering from byssinosis; the odds are $\exp(-2.668) = 0.069$ times lower of having the sickness in the medium and low dust levels as compared to the high level. The odds of having byssinosis increase with length of employment, with a parameter estimate of 0.625 (s.e. 0.170) or odds 1.87 times greater for more than ten years as opposed to less. Finally, non-smokers have a lower risk, with parameter estimate, -0.611 (s.e. 0.190) for odds 0.54 times smaller.

The analysis of these data resulted in a very simple and clearly interpretable model. Such simplicity will not always be the case in complex cross-sectional studies! However, there is also a danger that some important interaction might be missed, drowned in the global change in deviance when all interactions are simultaneously excluded, as we have done. (The reader might like to refer to the analysis by Aitkin *et al.*, 1989, pp. 205–

Table 2.10. Toxaemic signs exhibited by mothers during pregnancy, as related to smoking habits and social class. (Brown *et al.*, 1983)

		Hypertension			
		Yes		No	
Cigarettes	Social	Protein urea			
smoked	class	Yes	No	Yes	No
0	I	28	82	21	286
1–19		5	24	5	71
20+		1	3	0	13
0	II	50	266	34	785
1–19		13	92	17	284
20+		0	15	3	34
0	III	278	1101	164	3160
1–19		120	492	142	2300
20+		16	92	32	383
0	IV	63	213	52	656
1–19		35	129	46	649
20+		7	40	12	163
0	V	20	78	23	245
1–19		22	74	34	321
20+		7	14	4	65

213, where a different elimination procedure is used, resulting in a more complex model.)

2.6 Multivariate responses

In contrast to the binomial logistic model fitted by generalized linear model software, the log linear model can be applied to data where there are several responses. The general formula in Equation (2.9) can be applied directly.

Table 2.10 gives the results of a study on the toxaemic signs, hypertension (`HYPER`) and protein urea (`UREA`), exhibited by mothers during pregnancy for a first child in Bradford, England. These are related to the number of cigarettes smoked per day by the mother during pregnancy (`SMOKE`) and the social class of the mother (`SC`). Here, the two responses are hypertension and protein urea and the minimal model for independence is

$$\texttt{HYPER} + \texttt{UREA} + \texttt{SMOKE} * \texttt{SC}$$

This has a deviance of 672.85 (706.85) with 43 d.f. Because we expect the two responses to be interdependent, we first add this to the model:

$$\texttt{HYPER} * \texttt{UREA} + \texttt{SMOKE} * \texttt{SC}$$

Table 2.11. Parameter estimates from the model of Equation (2.12) for the toxaemia data of Table 2.10.

	Estimate	s.e.
UREA(2)·HYPER(2)	1.369	0.061
UREA(2)·SC(2)	0.450	0.167
UREA(2)·SC(3)	0.200	0.142
UREA(2)·SC(4)	0.037	0.155
UREA(2)·SC(5)	−0.145	0.171
HYPER(1)·SMOKE2(2)	−0.431	0.041

for a deviance of 179.03 (215.03) with 42 d.f. Looking at the four other simple dependencies, we discover that adding the dependence of hypertension on smoking,

$$\mathtt{HYPER} * \mathtt{UREA} + \mathtt{SMOKE} * \mathtt{SC} + \mathtt{HYPER} \cdot \mathtt{SMOKE}$$

gives a deviance of 69.49 (109.49) with 40 d.f. Putting in the dependence of protein urea on social class,

$$\mathtt{HYPER} * \mathtt{UREA} + \mathtt{SMOKE} * \mathtt{SC} + \mathtt{HYPER} \cdot \mathtt{SMOKE} + \mathtt{UREA} \cdot \mathtt{SC}$$

further reduces it to 46.12 (94.12) with 36 d.f. No other relationship reduces the deviance very much. However, when we inspect the parameter estimates, we observe that those for the second and third categories of smoking, as related to hypertension, are similar. Grouping them in a variable, SMOKE2, we have the model

$$\mathtt{HYPER} * \mathtt{UREA} + \mathtt{SMOKE} * \mathtt{SC} + \mathtt{HYPER} \cdot \mathtt{SMOKE2} + \mathtt{UREA} \cdot \mathtt{SC} \qquad (2.12)$$

with a deviance of 46.49 (92.49) with 37 d.f.

The parameter estimates for the interdependence of HYPER and UREA and their dependence on SC and SMOKE2 in this final model are given in Table 2.11. Urea protein and hypertension are positively associated. Given a hypertension category, the probability of not having urea protein decreases with increasing order of social class, except for Class I, which is about the same as Class IV. Given a urea protein category, the probability of hypertension is relatively higher among smokers.

Cook's distances, in Figure 2.9, show that the category (observation 20) with neither toxaemic sign in Class II smoking one to 19 cigarettes fits badly. This large frequency is overestimated by the model. The normal probability plot, in Figure 2.10, has several off-diagonal residuals at the two ends, observation 28 with a large positive value and 20 and 27 with negative values. These anomalies should be investigated, but the model generally fits well.

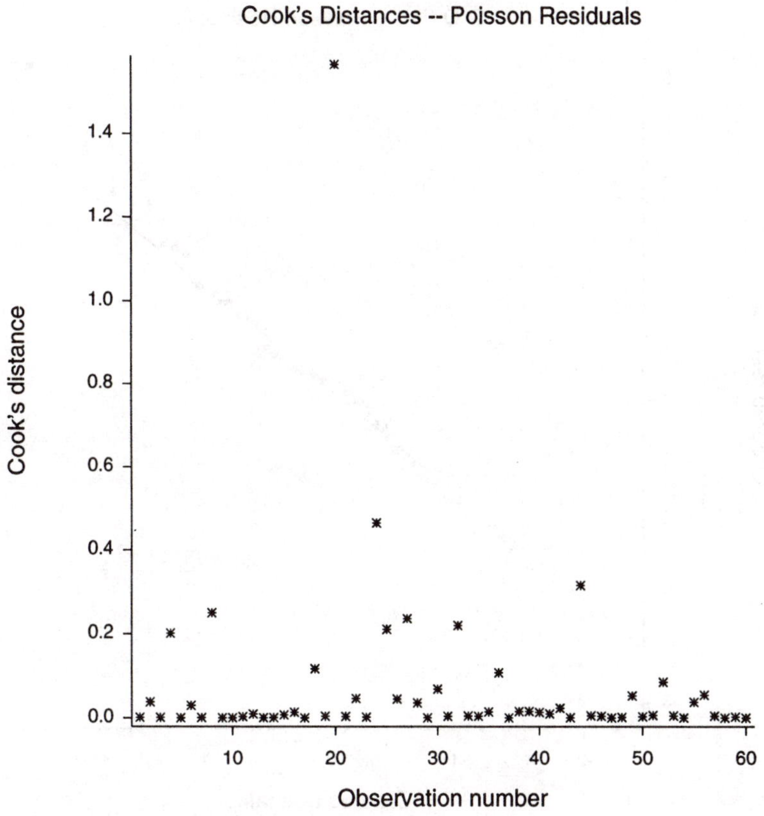

Fig. 2.9. Cook's distances from the model of Equation (2.12) for the toxaemia study of Table 2.10.

2.7 Genetic models

Frequencies of observed characteristics are often recorded in biological studies. In genetics, they can be predicted accurately following the theories of inheritance of such characteristics. For example, take Mendelian factors, where two possible phenotypes are observable. One is recessive (R) and the other dominant (D); the latter will appear more often than the former in the crossing of living things. With the interbreeding for two such factors, we expect to obtain four classes, $D_1D_2, D_1R_2, R_1D_2, R_1R_2$, in the ratio $9:3:3:1$ if the two factors segregate independently and all four classes of young are equally viable.

Fisher (1958, p. 82) gives data, shown in Table 2.12, for interbreeding of a plant (*Primula*), where the two factors concern the leaves, with normal being dominant, and flowers, with flat being dominant. A log linear model

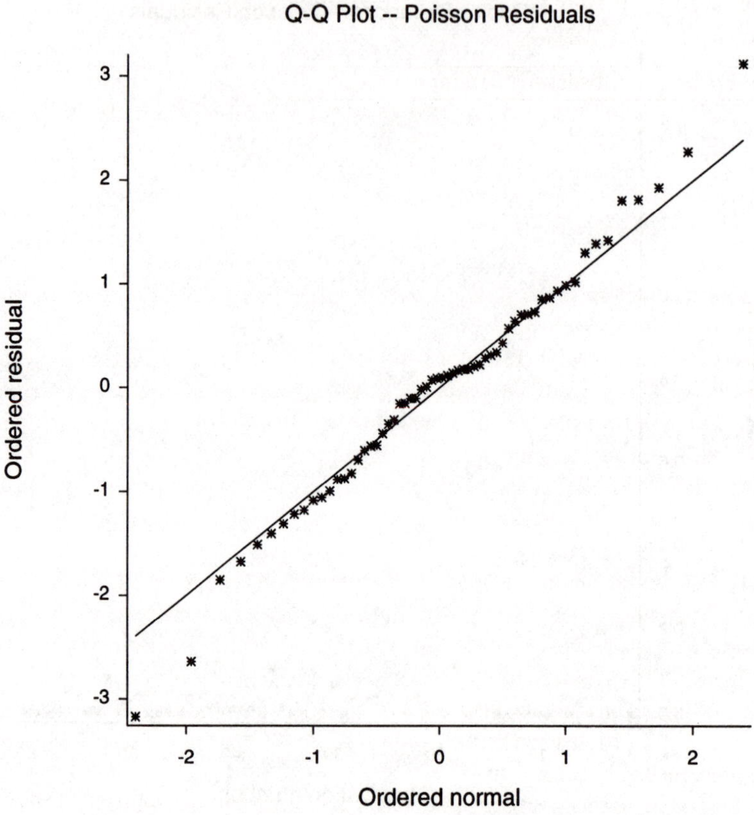

Fig. 2.10. Normal probability plot from the model of Equation (2.12) for the toxaemia study of Table 2.10.

Table 2.12. Results of an interbreeding experiment with *Primula*. (Fisher, 1958, p. 82)

	Leaves	
Flower	Flat	Crimped
Normal	328	77
Primrose	122	33

for independence between the two factors

$$\texttt{LEAVE} + \texttt{FLOWER}$$

fits the data very well, with a deviance of 0.36 (6.36) and 1 d.f. This result

Table 2.13. Fitted values from three models for the interbreeding experiment of Table 2.12.

Factors	Observed	Independence	9 : 3 : 3 : 1	Viability
NF	328	325.45	315	337.5
PF	122	124.55	105	112.5
NC	77	79.55	105	82.5
PC	33	30.45	35	27.5

indicates that there is no linkage between the two genes, i.e. they are not located on the same chromosome.

However, this model does not take into account the expected structure in the data. To do this, we want to fix the proportions in the ratios given above. For this, we place the logarithms of these values in a vector which is used as an offset. Then a null model is fitted, giving a deviance of 11.50 (13.50) with 3 d.f. Thus, our new model does not fit very well. The fitted values are given in Table 2.13, along with those for the independence model.

Fisher suggests that the plants with crimped leaves may be, to some extent, less viable than those with flat leaves. Because we have no indication of the degree of relative viability, we take the totals for each of the flat and crimped leaves as observed, and divide each of these classes in the ratio 3 : 1. This model can be fitted by changing the offset to these new ratios and putting the variable `LEAVE` back in the model. The resulting deviance is 2.46 (6.46) with 2 d.f., an acceptable fit. The fitted values are also given in Table 2.13.

If linkage had been found, the four probabilities are related by $\pi_{11} = (2+\alpha)/4$, $\pi_{12} = \pi_{21} = (1-\alpha)/4$, and $\pi_{22} = \alpha/4$, where $\sqrt{\alpha}$ is the recombination fraction (Fisher, 1958, pp. 300–306). The resulting model is non-linear and cannot be estimated as a log linear structure. This can be overcome simply by changing the link function to the identity. A vector derived from the constant terms in the above probabilities, (0.5, 0.25, 0.25, 0.0), multiplied by the total number of observations, is used as offset. A vector corresponding to the coefficient of α above, (0.25, −0.25, −0.25, 0.25), also multiplied by the total number of observations, is then fitted.

Fisher (1958, p. 299) gives data, reproduced in Table 2.14, on a study of linkage from the progeny of self-fertilized heterozygotes for two factors in maize. The independence model gives a deviance of 386.06 (392.06) with 1 d.f., while the fixed ratio model has 387.51 (389.51) with 3 d.f. The fitted values are given in Table 2.15. When the linkage model is fitted, with identity link, the deviance is 2.02 (4.02) with 3 d.f., indicating an excellent fit. We have the parameter estimate, $\hat{\alpha} = 0.0357$ (s.e. 0.005 84). Thus, the recombination fraction is estimated as 0.189.

With a little thought and judicious use of link functions, many theoret-

Table 2.14. Results of a self-fertilizing experiment with maize. (Fisher, 1958, p. 299)

	Base leaf	
Content	Green	White
Starchy	1997	906
Sugary	904	32

Table 2.15. Fitted values from three models for the self-fertilizing experiment of Table 2.14.

Factors	Observed	Independence	9 : 3 : 3 : 1	Linkage
StG	1997	2193.7	2159.44	1953.77
StW	906	709.3	719.81	925.48
SuG	904	707.3	719.81	925.48
SuW	32	228.7	239.94	34.27

ical models can be fitted to frequency data.

2.8 Exercises

(1) (a) Reanalyse the U.S.A. suicide data of Exercise (1.4) using the techniques of this chapter to find a seasonal trend model which fits all three years simultaneously.

(b) What are the comparative advantages of this simultaneous estimation and using an offset, as was suggested in Exercise (1.4)?

(2) In an experiment to study DDT, *Drosophila* fly larvae were reared in a medium containing that chemical. Pupae were classified by location and the resulting adults by sex and poisoning, as in the following table (Sokal and Rohlf, 1969, p. 602):

		Mortality	
Sex	Pupation site	Healthy	Poisoned
Male	In medium	55	6
Female		34	17
Male	Medium margin	23	1
Female		15	5
Male	Vial wall	7	4
Female		3	5
Male	Top of medium	8	3
Female		5	3

Determine if pupation site affects mortality and if this differs with sex.

(3) Breslow and Day (1982, p. 155) present data from the Ille-et-Vilaine, France, study of oesophageal cancer. Cases were 200 males diagnosed with this cancer in a regional hospital between January, 1972 and April, 1974. Controls were a sample drawn from the electoral list. Among other variables, information was collected on alcohol and tobacco consumption (g/day), as shown in the table:

	Tobacco			
Alcohol	0–9	10–19	20–29	30+
	Cases			
0–39	9	10	5	5
40–79	34	17	15	9
80–119	19	19	6	7
120+	16	12	7	10
	Controls			
0–39	252	74	35	23
40–79	145	68	47	20
80–119	42	30	10	5
120+	8	6	5	3

Study the joint dependence of oesophageal cancer on these two risk factors.

(4) A case-control study was designed to look at the dependence of arteriosclerosis on age, sex, and smoking behaviour (Aickin, 1983, p. 215):

	Age					
Smoking	35–54		55–64		65+	
behaviour	Case	Cont.	Case	Cont.	Case	Cont.
	Male					
Never smoked	3	23	3	15	12	15
Quit > 5 years	2	9	5	8	8	14
Quit < 5 years	3	11	1	1	3	2
< 1pack/day	4	9	3	2	14	4
1 pack/day	9	20	14	4	10	2
> 1 pack/day	20	24	34	10	15	3
	Female					
Never smoked	3	135	18	165	58	126
Quit > 5 years	1	13	3	15	7	18
Quit < 5 years	0	11	1	8	3	3
< 1 pack/day	12	42	21	28	18	11
1 pack/day	22	45	28	20	10	3
> 1 pack/day	14	16	19	15	8	2

Is there any evidence of such dependencies?

(5) Consider data on the presence of *torus mandibularis*, a protuberance

of the jaw, in Inuit tribes (Haberman, 1974a, p. 54, from Muller and Mayhall):

			Age					
T.m.	Sex	Tribe	1–10	11–20	21–30	31–40	41–50	> 50
Yes	M	Igloolik	4	8	13	18	10	12
No			44	32	21	5	0	1
Yes	F		1	11	19	13	6	10
No			42	17	17	5	4	2
Yes	M	Hall	2	5	7	5	4	4
No		Beach	17	10	6	2	2	1
Yes	F		1	3	2	5	4	2
No			12	16	6	2	0	0
Yes	M	Aleut	4	2	4	7	4	3
No			6	13	3	3	5	3
Yes	F		3	1	2	2	2	4
No			10	7	12	5	2	1

Is there any evidence that this condition varies in different subgroups of the population?

(6) Psychiatric patients were cross-classified by their symptoms (Wermuth, 1976):

		Depression			
		Yes		No	
		Stability			
Solidity	Validity	Intro.	Extro.	Intro.	Extro.
Rigid	Energetic	15	23	25	14
Hysteric		9	14	46	47
Rigid	Psychasthenic	30	22	22	8
Hysteric		32	16	27	12

(The stability categories are introvert and extrovert.) What relationships of dependence exist among these four symptoms?

(7) Twin births were reported in six cities between October 1961 and December 1964 (Andersen, 1980, p. 93):

City	2 boys	2 girls	1 each
Alexandria	116	114	161
Hong Kong	45	46	34
Melbourne	29	36	33
Santiago	88	77	76
Sao Paulo	61	69	81
Zagreb	20	32	30

(a) Are the probabilities of boys and girls being born equal?

(b) In a simple model, we would expect the probability of twins with a child of each sex to be twice as large as that for twins of identical sex. Does this model hold for these data?

(c) If not, develop a more complex model to describe the data?

(d) Is there any evidence of difference among the cities?

(8) The table below gives the sources of knowledge of cancer (Lombard and Doering, 1947, from Potter).

			Reading			
			Yes		No	
			Knowledge			
Lectures	Newspaper	Radio	Good	Poor	Good	Poor
Yes	Yes	Yes	23	8	8	4
No			102	67	35	59
Yes	No		1	3	4	3
No			16	16	13	50
Yes	Yes	No	27	18	7	6
No			201	177	75	156
Yes	No		3	8	2	10
No			67	83	84	393

Which media appear to have been the most beneficial?

(9) The following table presents results from a study of the health of British coal miners who were smokers but did not show radiological signs of pneumoconiosis (Ashford and Sowden, 1970).

	Breathless			
	Yes		No	
	Wheeze			
Age	Yes	No	Yes	No
20-24	9	7	95	1841
25-29	23	9	105	1654
30-34	54	19	177	1863
35-39	121	48	257	2357
40-44	169	54	273	1778
45-49	269	88	324	1712
50-54	404	117	245	1324
55-59	406	152	225	967
60-64	372	106	132	526

Study how breathlessness and wheeze simultaneously vary with age.

(10) The following table shows the educational aspirations of students as it relates to degree of parental encouragement, social class, and IQ (Sewell and Shah, 1968):

Parents encourage	College plans	IQ	Sex	Social class: Low	Low middle	Middle	High
Low	Y	L	M	4	2	8	4
High				13	27	47	39
Low	N			349	232	166	48
High				64	84	91	57
Low	Y	LM		9	7	6	5
High				33	64	74	123
Low	N			207	201	120	47
High				72	95	110	90
Low	Y	UM		12	12	17	9
High				38	93	148	224
Low	N			126	115	92	41
High				54	92	100	65
Low	Y	H		10	17	6	8
High				49	119	198	414
Low	N			67	79	42	17
High				43	59	73	54
Low	Y	L	F	5	11	7	6
High				9	29	36	36
Low	N			454	285	163	50
High				44	61	72	58
Low	Y	LM		5	19	13	5
High				14	47	75	110
Low	N			312	236	193	70
High				47	88	90	76
Low	Y	UM		8	12	12	12
High				20	62	91	230
Low	N			216	164	174	48
High				35	85	100	81
Low	Y	H		13	15	20	13
High				28	72	142	360
Low	N			96	113	81	49
High				24	50	77	98

Do student college plans depend systematically on any of these family and individual characteristics?

(11) Data were collected in a Czechoslovakian car factory at the beginning of a 15-year prospective study of possible risk factors for coronary thrombosis. The following table cross-classifies several prognostic factors for coronary heart disease (Edwards and Havranek, 1985, from Reinis *et al.*).

Strenuous physical work	Systolic blood pressure	Ratio of beta/alpha lipoprotein	Family heart disease	Strenuous mental work: No, Smoking: No	No, Smoking: Yes	Yes, Smoking: No	Yes, Smoking: Yes
No	< 140	< 3	No	44	40	112	67
Yes				129	145	12	23
No	> 140			35	12	80	33
Yes				109	67	7	9
No	< 140	> 3		23	32	70	66
Yes				50	80	7	13
No	> 140			24	25	73	57
Yes				51	63	7	16
No	< 140	< 3	Yes	5	7	21	9
Yes				9	17	1	4
No	> 140			4	3	11	8
Yes				14	17	5	2
No	< 140	> 3		7	3	14	14
Yes				9	16	2	3
No	> 140			4	0	13	11
Yes				5	14	4	4

Which factors are most closely related?

(12) In his classical genetic experiment,, Mendel crossed smooth yellow peas with wrinkled green peas, the former being dominant for each factor. The results were

Shape	Colour: Yellow	Colour: Green
Round	315	108
Wrinkled	101	32

(a) Construct a model for these data.
(b) Do the results indicate anything unusual about the data?

3
Regression models

The general approach to entering explanatory variables into a model is through a regression. This is effectively what we were doing in Chapter 1, having the frequencies depend on time, season, etc. Nominal explanatory variables are a special case. Normally, a set of zero–one indicator variables would have to be specified for the different categories of the variable. However, generalized linear model software generally handles this automatically, by factor variables, as is also indicated by the notation which we have been using. In this chapter, we shall investigate some more complex models containing quantitative explanatory variables, which one may not want to include as categorical variables.

3.1 Time trends

In Chapter 1, we studied how recall of an event varied with time. There, we had a simple frequency table. We shall now consider a more complex case, where we have a vector of responses at each time point. Our example concerns the `ATTITUDE` towards treatment of criminals by the courts in the U.S.A. between 1972 and 1975 (`YEAR`), as given in Table 3.1. We have the number of responses to each category over a period of five years. However, this is a series of four cross-sectional studies, not a panel, because the same individuals were not involved each time. Thus, we have no indication of how individual opinions change. Models for panel data will be considered in Part II.

Table 3.1. Changes in attitudes to criminals, 1972–1975. (Haberman, 1978, p. 128)

	1972	1973	1974	1975
Too Harshly	105	68	42	61
Not Harshly Enough	1066	1092	580	1174
About Right	265	196	72	144
Don't Know	173	138	51	104
No Answer	4	10	8	7
Total	1613	1504	753	1490

We are interested in time trends in opinion, but relative to other categories of opinion. Preliminary inspection of Table 3.1 reveals that frequencies in all categories except 'not harshly enough' seem to be decreasing with time. (Note that, as can be seen from the marginal column totals, sample size for 1974 is about half that for the other years.) Thus, increase in one category must be studied in relation to a corresponding reduction of others. This is exactly what our log linear model does, because marginal totals are fixed with this model.

As usual, we first look at independence between attitude and year — that attitudes do not change over the years:

$$\mathtt{ATTITUDE} + \mathtt{YEAR}$$

The model is highly implausible, with a deviance of 87.05 (103.05) and 12 d.f.; we must reject independence. Attitudes do change over the years.

We shall now fit a linear trend. We introduce the dependence of attitude on years (YEARL) into our model as

$$\mathtt{ATTITUDE} + \mathtt{YEAR} + \mathtt{ATTITUDE} \cdot \mathtt{YEARL} \tag{3.1}$$

In this case, the linear effect of the year is not included among the main effects, but only as an interaction. This model will allow each attitude to have a different (log) linear change over time, with all comparisons to the slope of the last attitude.

Here, 'no answer' seems to be a reasonable baseline category, because one might hope that it should be relatively stable over time. The parameter values for the interaction are (−1.468, −0.513, −1.750, −1.579, 0.000). We discover that, in relation to this last attitude, the other four all decrease with time, but the second attitude, that criminals are not treated harshly enough, decreases less than the other three. The deviance is 13.87 (37.87) with 8 d.f., indicating a fairly reasonable fit.

A close inspection of the residuals for the independence model (not shown) would already have indicated that the observations for the attitude 'not harshly enough' fitted that model most poorly. This is still evident in the graph of Figure 3.1 giving Cook's distances for the linear trend model of Equation (3.1).

Instead of attempting to fit a more complex time trend, let us then eliminate this attitude 'not harshly enough' from the model by giving it zero weight in order to see if the remaining attitudes are independent of time, i.e. all change in the same way over time. The deviance of 16.58 (38.58) with 9 d.f. is borderline, so that all attitudes except 'not harshly enough' do not appear to change with time. Inspection of the residuals for this model indicates, however, that the 'no answer' response may also vary with time as compared to the others. Elimination of this category, by

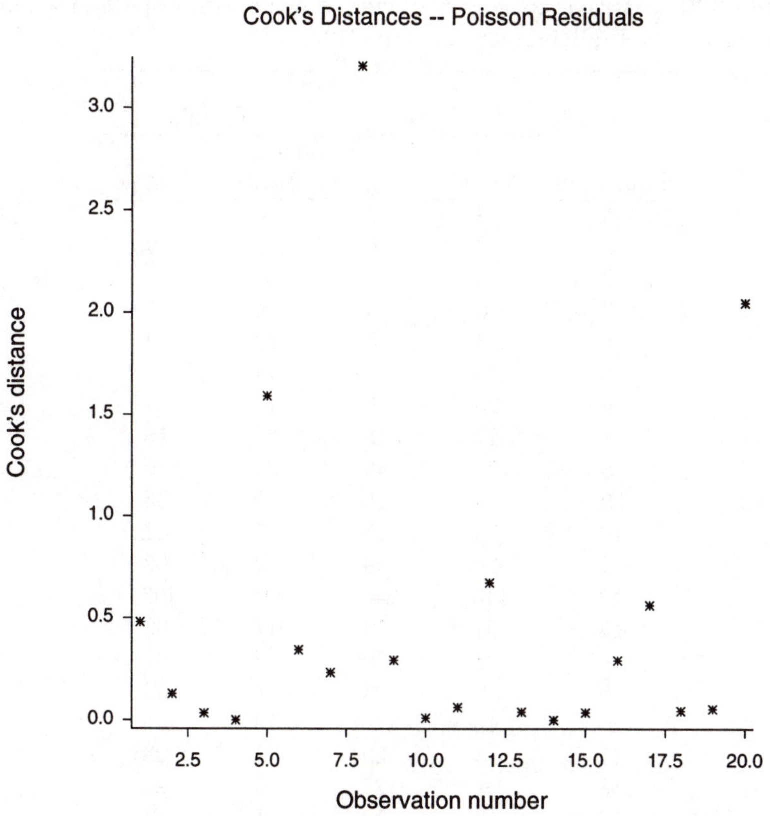

Fig. 3.1. Cook's distances from the linear time trend model of Equation (3.1) for the attitude to criminals data of Table 3.1.

weighting it out as well, improves the fit and leaves an acceptable model with a deviance of 5.77 (33.77) and 6 d.f. The categories (1) 'too harshly', (3) 'about right', and (4) 'don't know' apparently do not vary, with respect to each other, over the four years. Inspection of the residuals and plots (not shown) indicates no systematic patterns.

Our two approaches are complementary. Introduction of a linear time trend indicates how certain attitudes change, while elimination of certain attitude categories allows a check isolating those attitudes which have not changed among themselves. We conclude that the attitudes 'too harshly', 'about right', and 'don't know' remain relatively stable in relation to each other and are all losing ground to the 'not harshly enough' attitude.

Table 3.2. Attitude to women staying at home with respect to sex and educational level. (Haberman, 1979, p. 312)

	Sex			
	Male		Female	
	Attitude			
Education	Agree	Disagree	Agree	Disagree
0	4	2	4	2
1	2	0	1	0
2	4	0	0	0
3	6	3	6	1
4	5	5	10	0
5	13	7	14	7
6	25	9	17	5
7	27	15	26	16
8	75	49	91	36
9	29	29	30	35
10	32	45	55	67
11	36	59	50	62
12	115	245	190	403
13	31	70	17	92
14	28	79	18	81
15	9	23	7	34
16	15	110	13	115
17	3	29	3	28
18	1	28	0	21
19	2	13	1	2
20	3	20	2	4

3.2 Model simplification

Any continuous variable may always be treated as a nominal variable; the values can only be measured to some finite precision and order can be ignored. This, however, can involve several disadvantages. A large number of categories may be required to represent the relationship adequately, with a correspondingly large number of parameters in the model. At the same time, a nominal representation does not exploit the structure of the data as fully as is possible. Models may be simplified and interpretation aided by the use of continuous variables.

Consider an example concerning the study of `ATTITUDE` towards women staying at home (agree/disagree) as it depends on `EDUCATION` and `SEX`, given in Table 3.2. Here, the education variable has 21 categories corresponding to years of study, giving a table with 84 frequencies and, hence, the possibility of a model with as many parameters (or half as many using a logistic model).

With a binary response, we use the binomial distribution. (Care must be taken with some software because there are no observations for women at education level 2.) The null model, with only a general mean, is highly implausible, having a deviance of 451.72 (453.72) with 40 d.f. Attitudes to women staying at home depend either on sex or on education or on both.

A model with only sex as the explanatory variable must also be rejected: the deviance is 451.71 (455.71) with 39 d.f. At this point, differences by sex appear not to be important, because the fit is not improved over the independence model.

Introduction of a nominal education variable in place of sex explains a lot of the variability. With a deviance of 27.66 (69.66) and 20 d.f., it leaves little lack of fit, but we have a complex model with 21 parameters. From the parameter estimates (not shown), we see that, in general, it is more probable for those with lower education levels to be favourable to women staying at home. In the binomial model, we are studying the relation agree/disagree. The larger parameter estimates at low education levels than at high indicate more chance of agreeing than disagreeing at these low levels as compared to the higher education levels.

We shall now try a linear trend variable for education completed (EDUCL). Although the model is greatly improved over the null model, the deviance for lack of fit with respect to the saturated model, 64.03 (68.03) with 39 d.f., is large. From the negative parameter estimate for the slope, −10.62, we now see more clearly that agreeing that women should stay at home decreases with increasing education: the ratio of the number of people agreeing to disagreeing decreases as education increases.

We may next look at interaction between sex and the linear effect of education. To do this, we must put sex back into the model to keep it hierarchical, giving a model represented as

$$\texttt{EDUCL} * \texttt{SEX} \tag{3.2}$$

With a deviance of 57.10 (65.10) and 37 d.f., the model is better but still perhaps not sufficiently good. We may note that agreement with women staying at home is about the same, on average, for the two sexes (−0.045), that it decreases with education (−10.78), but that it decreases less quickly for men ($-10.78 + 1.597 = -9.68$) than for women ($-10.78 - 1.597 = -12.38$): the interaction effect.

If we add the quadratic main effect and its interaction with sex, the deviance virtually does not change at all. If we want a simple model, we seem to be left with one which may not fit the data sufficiently well. However, if we look at Cook's distances for our model, in Figure 3.2, we see that observation 18 (female education level 8) fits the data less well than the others. This might be what is known as an outlier. We may have

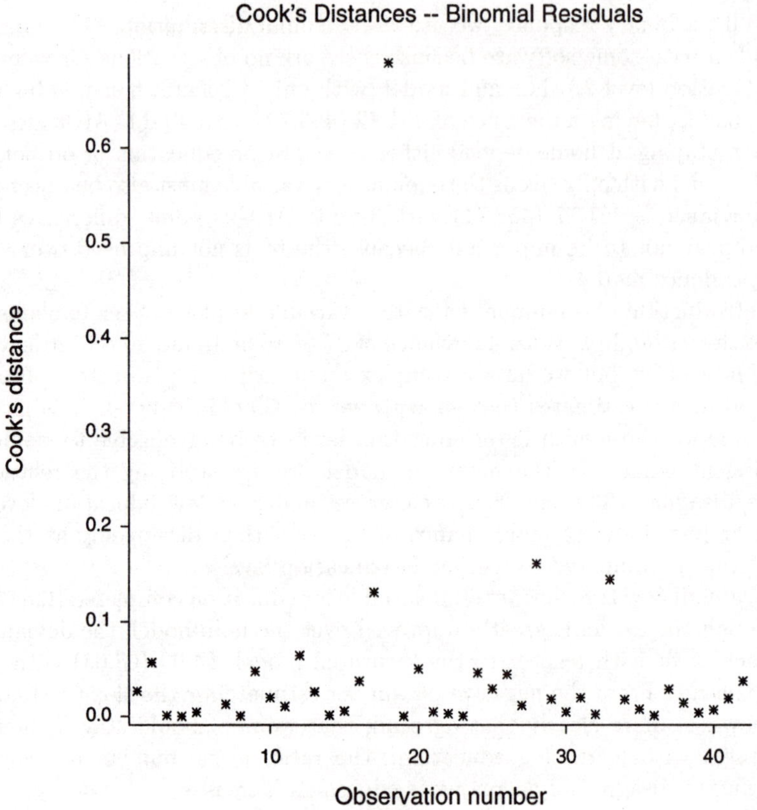

Fig. 3.2. Cook's distances from the linear trend model of Equation (3.2) for the women staying at home study of Table 2.8.

an anomaly in the data. Here, with large frequencies in these categories, it is not likely to be an error in the original coding for several individual women, but might rather be a mistake in recording the frequency itself. Unfortunately, it is not possible to check this in the secondary analysis of data which we carry out here.

Study of the list of residuals for our linear model (not shown) would also show that the large residuals are primarily for education levels less than six. This is not obvious from Cook's distances in Figure 3.2, but it is reflected in the bends at the two ends of the normal probability plot in Figure 3.3. We may try eliminating these lower levels from our model by weighting them out (which is equivalent to fitting a different factor level to each). The deviance is 36.02 (66.02) with 26 d.f. We have a simple model which fits reasonably well. The parameter estimates have changed

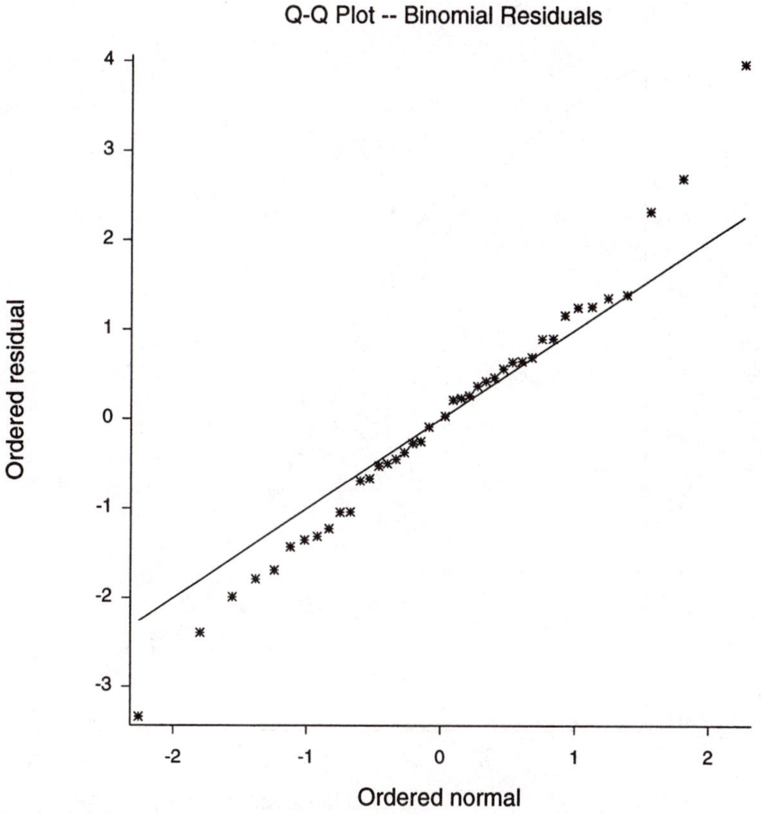

Fig. 3.3. Normal probability plot from the linear trend model of Equation (3.2) for the women staying at home study of Table 2.8.

very little from the previous model. Our conclusions above, that agreement with women staying at home decreases with education (above five years), but less quickly for men than for women, still holds. A last graph, Figure 3.4, shows how fitted and observed values change with educational level. Note that estimation of these fitted curves did not involve the observed values under six years of education.

This graph presents two curves, one for men and another for women, with the corresponding observed values around them. Such curves of fitted values are known as logistic curves. For our data, they decrease because the estimated slopes are negative. Notice how they flatten off on top at one without quite reaching it and the same at the bottom before reaching zero. The curve for men starts off lower than that for women and ends higher, i.e. is flatter. The contrast between low and high education is greater for

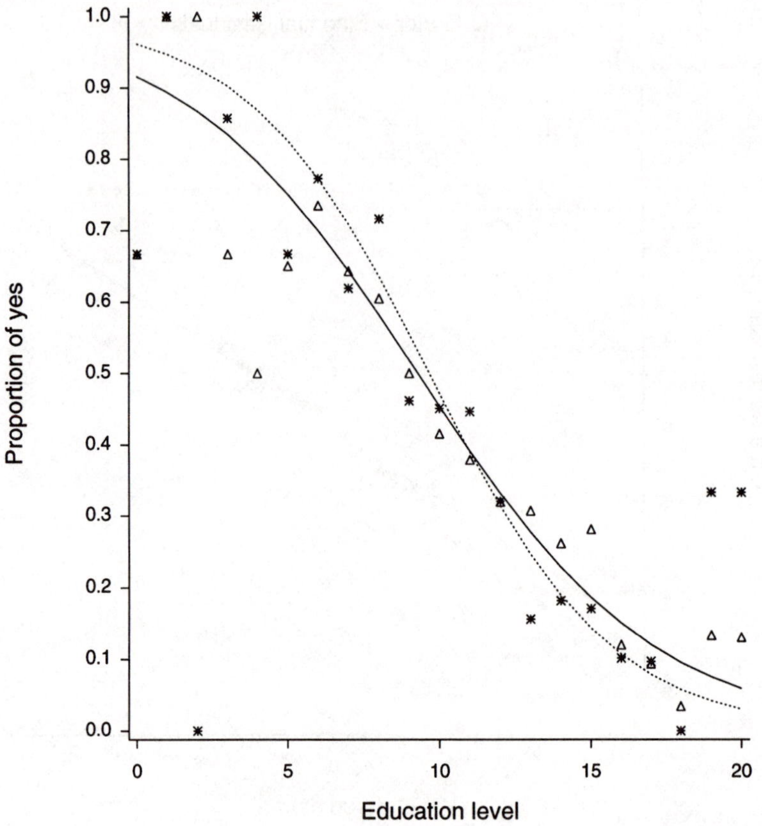

Fig. 3.4. Observed values and fitted model of Equation (3.2) for the data of Table 3.2 on women staying home (triangles and solid line: men; stars and dotted line: women).

women than for men, as was already indicated by the parameter values.

3.3 Quantal response models

A biological assay, or bioassay, is an experiment to study the reaction of a living organism to some material. When the only response recorded is presence or absence of the reaction, it is called a quantal response. Usually, the material to be studied is applied at different dose levels and the response observed. In many such assays, the material may cause irreversible change, such as death, so that the subjects can only be used once.

One of the very early quantal response assays, published in 1926, concerned a study of the effect of insulin on mice, as measured by whether or not they had convulsions. The data are given in Table 3.3. We are

Table 3.3. Number of mice with convulsions under various doses (0.001 IU) of insulin and two preparations. (Finney, 1978, p. 376, from Hemmingsen and Krogh)

Standard preparation			Test preparation		
Dose	With convulsions	Total	Dose	With convulsions	Total
3.4	0	33	6.5	2	40
5.2	5	32	10.0	10	30
7.0	11	38	14.0	18	40
8.5	14	37	21.5	21	35
10.5	18	40	29.0	27	37
13.0	21	37			
18.0	23	31			
21.0	30	37			
28.0	27	30			

interested in modelling how the proportion of mice with convulsions varies with the dose of insulin and with the type of preparation.

The idea behind such a model is that each organism reacts only to dose levels greater than some minimum, different for each organism. Then, these threshold levels have a distribution in the population under study whose cumulative distribution function is $\pi = F(\beta_0 + \beta_1 x)$, where x is the dose.

A simple model for these data is logistic regression, which allows the logit of the probability of convulsions to depend linearly on dose and on the preparation:

$$\log\left(\frac{\pi_i}{1-\pi_i}\right) = \beta_0 + \beta_1 x_{1i} + \beta_2 x_{2i}$$

where π_i is the probability of convulsions at dose, x_{1i}, with preparation, x_{2i} (coded 0 for standard and 1 for test preparation). This corresponds to a logistic distribution for the threshold levels, with

$$\pi_i = \frac{e^{\beta_0+\beta_1 x_{1i}+\beta_2 x_{2i}}}{1+e^{\beta_0+\beta_1 x_{1i}+\beta_2 x_{2i}}}$$

the logistic distribution function.

Many other models are also possible. For example, Finney (1978) uses a probit link function, corresponding to a normal cumulative distribution function, although this will generally give similar results to the logit. In that case, the normal distribution describes the reaction levels. Note that both the normal and logistic distributions are symmetric.

Another possible data generating mechanism for quantal response data can be described by the multi-hit model, whereby the organism is thought to be bombarded by the dose, but requires one or more 'hits' to react. Closely related to this, mathematically, is the multi-stage model in which the organism must pass through several stages before reacting. If the hits are arriving as a Poisson process, i.e. the lengths of the stages have exponential distributions, then the distribution function will be gamma (Morgan, 1992, pp. 22–25).

With the binomial distribution and logit link, the model is fitted as

DOSE + PREP

The estimates of the parameters are $\hat{\beta}_0 = -2.078$, $\hat{\beta}_1 = 0.161$, and $\hat{\beta}_2 = -0.875$. Because the slope for dose is estimated to be positive, the proportion of mice having convulsions increases with dose. The coefficient for type of preparation is estimated to be negative, indicating fewer convulsions with the second, the test preparation.

The deviance of the model is 27.10 (33.10) with 11 d.f., indicating that it is not very acceptable. The increase in deviance when dose is omitted from the model ($\beta_1 = 0$) is 138.81 (AIC 169.91) and when preparation is removed ($\beta_2 = 0$), 14.91 (AIC 46.01), both with 1 d.f., indicating that both parameters are necessary to the model. One possible defect of the model is that convulsions may depend on dose in a different way for the two preparations. An interaction term, the product of the two explanatory variables, $x_{1i}x_{2i}$, may be added to the model,

DOSE + PREP + DOSE · PREP

with a decrease in deviance of 4.23 and 1 d.f. The deviance is 22.87 (30.87) with 10 d.f., some improvement. The estimates of the parameters are now $\hat{\beta}_0 = -2.444$, $\hat{\beta}_1 = 0.193$, $\hat{\beta}_2 = 0.061$, and $\hat{\beta}_3 = -0.066$. Thus, the proportion of convulsions under the test preparation is now higher for very low doses (β_2, the difference in intercept, is positive), but increases less rapidly with dose than for the standard preparation (β_3, the difference in slope, is negative).

One possible improvement would be to fit a higher order polynomial in dose. In our case, a simpler approach is to try a transformation of the dose, such as the logarithm. The new linear regression model will be

$$\log\left(\frac{\pi_i}{1-\pi_i}\right) = \beta_0 + \beta_1 \log(x_{1i}) + \beta_2 x_{2i} \tag{3.3}$$

fitted as

LOGDOSE + PREP

This corresponds to a distribution of threshold levels which is log logistic

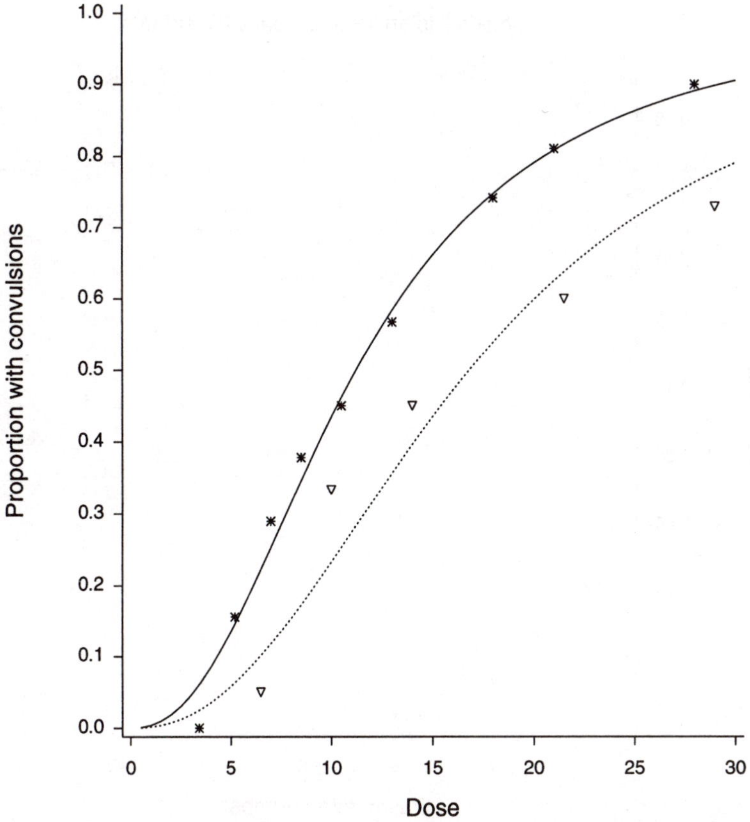

Fig. 3.5. Logistic regression model of Equation (3.3) for the insulin data of Table 3.3 (stars and solid line: observed values and fitted curve for the standard preparation; triangles and dotted line: the test).

and which is thus no longer symmetric. (If the probit link were used, we would have an underlying log normal distribution.) Such a model is often more reasonable for such phenomena. The estimates are now $\hat{\beta}_0 \doteq -5.553$, $\hat{\beta}_1 = 2.297$, and $\hat{\beta}_2 = -0.929$, and the deviance is 8.79 (14.79) with 11 d.f., an acceptable model. The conclusions are similar to those described above for the no interaction model. The response curves are plotted in Figure 3.5. As expected from the parameter values, the curve for the standard preparation climbs higher than that for the test. The former seems to fit more closely to the data. This is clearly reflected in the plot of Cook's distances in Figure 3.6. Note, however, that the values on the ordinate of this graph are smaller than those of most other such graphs which we have looked at. Although the deviance indicates a globally satisfactory fit, there

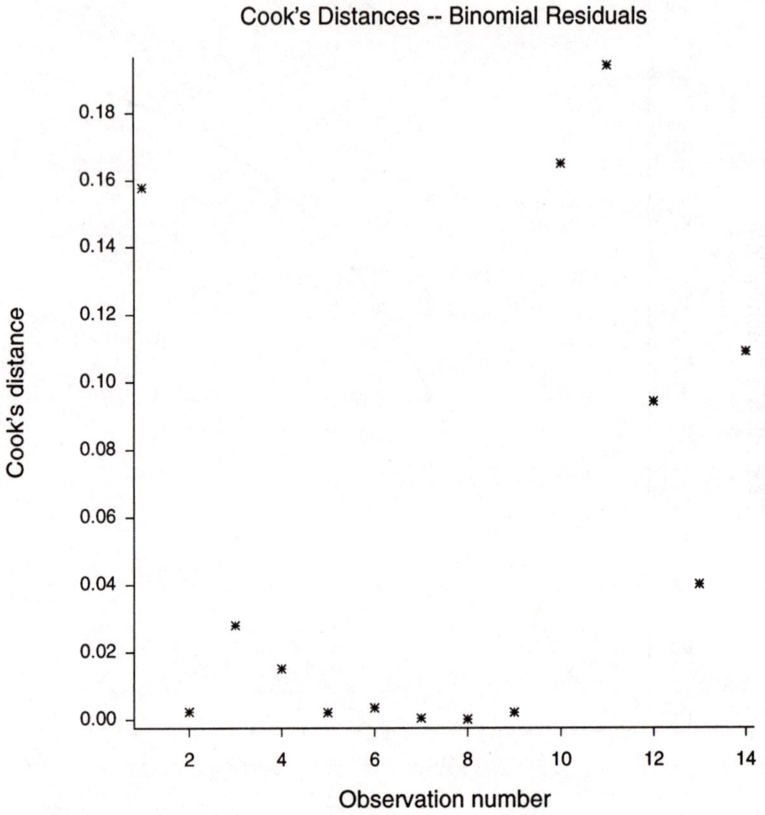

Fig. 3.6. Cook's distances from the logistic regression model of Equation (3.3) for the insulin data of Table 3.3.

is some indication that the shape of the curve for the test treatment is not logistic in log dose. This will be investigated further.

A reasonable possibility is to change the link function. The complementary log–log link, $\log[-\log(\pi)]$, can provide a steeper curve, as appears to be required here. Note that this link is not symmetric in successes and failures of the trials; it corresponds to an extreme value distribution for the thresholds. With convulsions as the response variable in the model, the deviance for the model of Equation (3.3) is 12.87 (18.87) with 11 d.f., considerably poorer than the logistic model. However, with no convulsions as response,

$$\log[-\log(1-\pi_i)] = \beta_0 + \beta_1 \log(x_{1i}) + \beta_2 x_{2i} \qquad (3.4)$$

the deviance is 4.69 (10.69) with 11 d.f., considerably better than the lo-

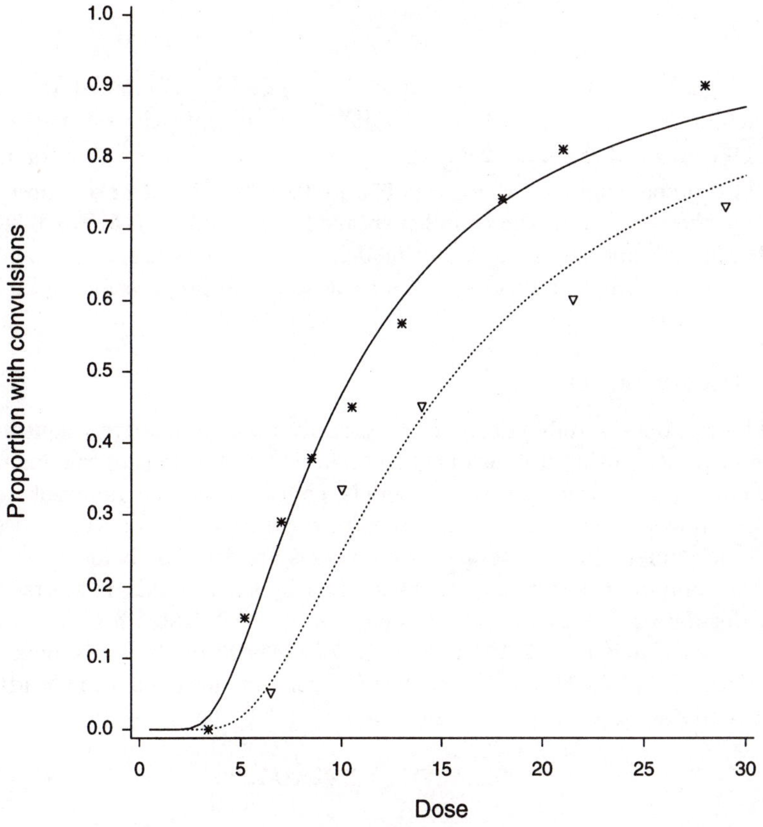

Fig. 3.7. Complementary log–log regression model of Equation (3.4) for the insulin data of Table 3.3 (stars and solid line for the observed values and fitted curve: the standard preparation; triangles and dotted line: the test).

gistic model. The estimates are $\hat{\beta}_0 = 3.286$, $\hat{\beta}_1 = -1.547$, and $\hat{\beta}_2 = 0.6099$ and the estimated curves are plotted in Figure 3.7. Although the fit for the standard preparation appears to be slightly poorer, that for the test is improved.

Often in quantal response models, a quantity of interest is the (log) dose which corresponds to $100p\%$ reaction, called the ED_{100p}. Thus, the ED_{50} is the median of the threshold distribution. For a simple regression model, we have

$$\beta_0 + \beta_1 \mathrm{ED}_{100p} = g(p)$$

where $g(\cdot)$ is the link function used, from which

$$\mathrm{ED}_{100p} = \frac{1}{\beta_1}[g(p) - \beta_0]$$

For a logit link, $g(0.5) = 0$, so that $\mathrm{ED}_{50} = -\beta_0/\beta_1$. For our data with the model of Equation (3.3) and this link, we calculate the estimates to be $\widehat{\mathrm{ED}}_{50} = 5.553/2.297 = 2.42$, corresponding to a dose of 11.2 for the standard preparation and $\widehat{\mathrm{ED}}_{50} = (5.553+0.929)/2.297 = 2.82$ to a dose of 16.8 for the test. With the complementary log–log link, $g(0.5) = -0.367$. With this link and the same linear model, the estimates are, respectively, 2.36 corresponding to a dose of 10.6 for the standard preparation and 2.76 to a dose of 15.7 for the test.

3.4 Binary regression

The response variable need not necessarily be a cumulative frequency, as we have been using up until now. Individual observations of raw binary data may also be used directly in models, coded as zero for no event and one for an event. For binary data, a corresponding vector of ones is used for the binomial denominators and the models are fitted as before.

One common event which can be studied by such models is mortality. Consider data on deaths in car accidents, as given in Table 3.4. The `AGE` of the person and the `VELOCITY` and `ACCELERATION` of the body making impact on the side of the car were also recorded in this simulation study.

A model with the three variables,

$$\texttt{AGE} + \texttt{VELOCITY} + \texttt{ACCELERATION} \tag{3.5}$$

has a deviance of 43.98 (51.98) with 54 d.f. We might conclude that this model fits very well. However, in contrast to models for larger frequencies, the absolute deviance for a binary data problem has no interpretation for goodness of fit, because it depends only on the fitted values. We can only look at changes in deviance when variables are added or removed from the model.

If we inspect the standard errors of the estimates, given in Table 3.5, we suspect that `ACCELERATION` can be removed from the model, and, indeed, the change in deviance is then only 1.36 (AIC 51.33). `VELOCITY` might be removed instead, but the fit is slightly poorer. Eliminating `AGE` or both `ACCELERATION` and `VELOCITY` greatly increases the deviance. The introduction of interactions and squared terms can also be studied, but these reduce the deviance very little. Thus, we are left with the model,

$$\texttt{AGE} + \texttt{VELOCITY} \tag{3.6}$$

The usual residual plots are given in Figures 3.8 and 3.9. Except for a couple of points (observations 23 and 35 — the two largest Cook's dis-

Table 3.4. Data on mortality due to simulated side impact car collisions (1 indicates mortality), as it depends on age, car velocity, and acceleration (units not specified). (Härdle and Stoker, 1989)

Age	Vel.	Acc.	Mort.	Age	Vel.	Acc.	Mort.
22	50	98	0	30	45	95	0
21	49	160	0	27	46	96	1
40	50	134	1	25	44	106	0
43	50	142	1	53	44	86	1
23	51	118	0	64	45	65	1
58	51	143	1	54	45	103	0
29	51	77	0	41	45	102	1
29	51	184	0	36	45	108	1
47	51	100	1	27	45	140	0
39	51	188	1	45	45	94	1
22	50	162	0	49	40	77	0
52	51	151	1	24	40	101	0
28	50	181	1	65	40	82	1
42	50	158	1	63	51	169	1
59	51	168	1	26	40	82	0
28	41	128	0	60	45	83	1
23	61	268	1	47	45	103	1
38	41	76	0	59	44	104	1
50	61	185	1	26	44	139	0
28	41	58	0	31	45	128	1
40	61	190	1	47	46	138	1
32	50	94	0	41	45	102	0
53	47	131	0	25	44	90	0
44	50	120	1	50	44	88	1
38	51	107	1	53	50	128	1
36	50	97	0	62	50	136	1
33	53	138	1	23	50	108	0
51	41	68	1	27	60	176	1
60	42	78	1	19	60	191	0

Table 3.5. Parameter estimates from the model of Equation (3.5) for the collision data of Table 3.4.

	Estimate	s.e.
1	−15.050	5.265
AGE	0.172	0.043
VELOCITY	0.146	0.112
ACCELERATION	0.016	0.014

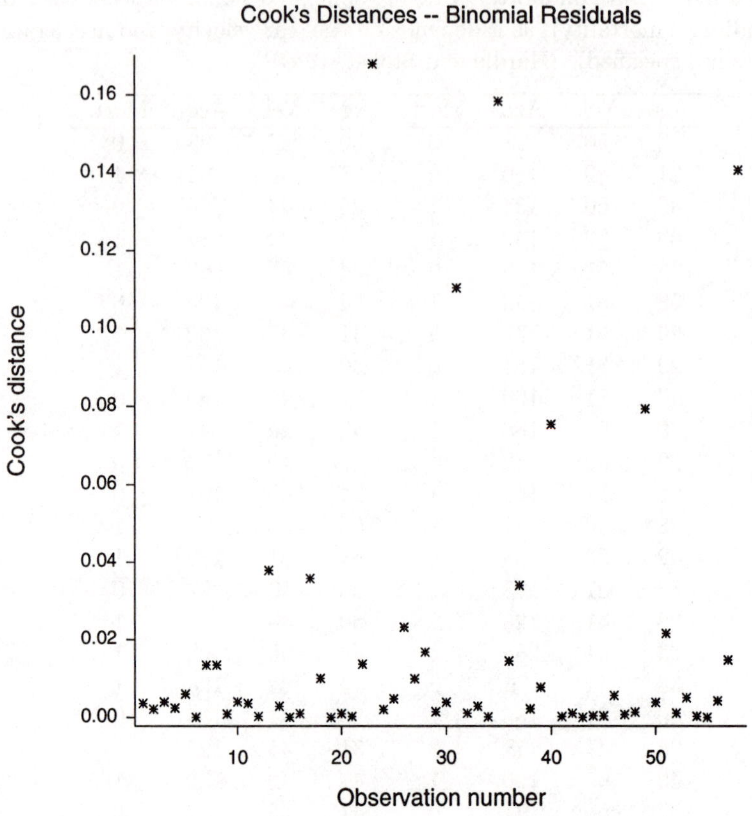

Fig. 3.8. Cook's distances from the model of Equation (3.6) for the car collision data of Table 3.4.

tances) in the lower left corner of the normal probability plot, these appear satisfactory.

The model is plotted in Figure 3.10 for `VELOCITY` with `AGE` fixed at 25, 40, and 55, and in Figure 3.11 for `AGE` with `VELOCITY` fixed at 45, 50, and 55. We see how mortality increases rapidly with age, levelling off at about 50, and with velocity, although the latter varies greatly with age.

3.5 Polytomous regression

The raw data of individual responses may also be modelled directly when they are polytomous. This will usually only be necessary when one or more of the explanatory variables are continuous, so that a cross-classified table cannot be set up without great loss of information.

The procedure is a direct extension of the method used to analyse binary

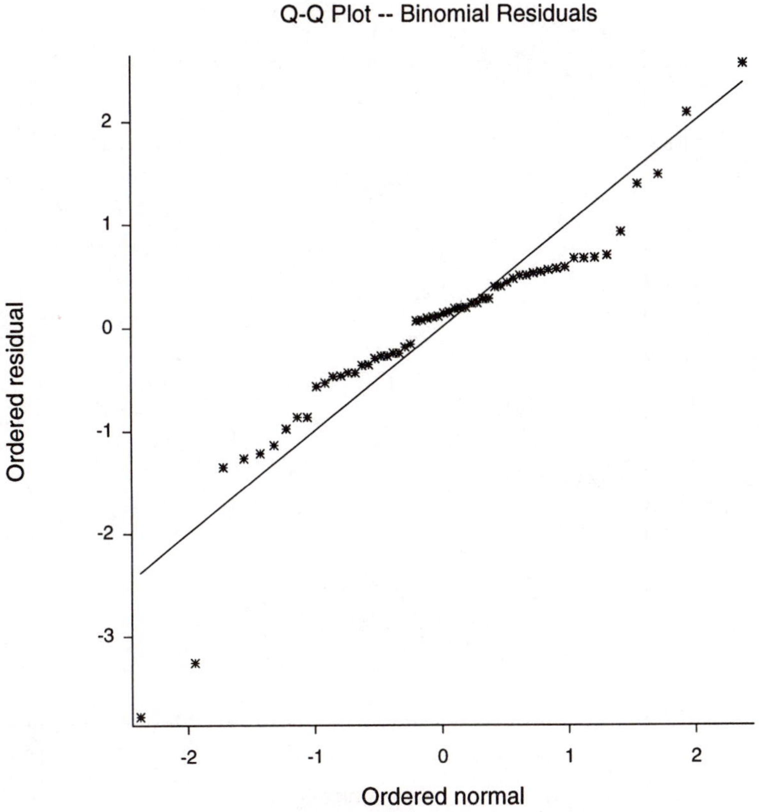

Fig. 3.9. Normal probability plot from the model of Equation (3.6) for the car collision data of Table 3.4.

responses as log linear models. As we have seen, this procedure results in the same model as if a logistic model were used. The sets of frequencies for the two possible responses were placed in the same vector, one set after the other, making the vector twice the length of that for the equivalent logistic regression. Likewise, when there are I categories of response, all vectors will be of length I times the number of observations. Thus, all explanatory variables have their values repeated I times. The 'response' vector for the Poisson distribution will have I sections. In the ith section, there will be a 1 if the individual had that response and a 0 otherwise. The total number of ones in the vector corresponds to the number of observations. Finally, a factor variable is constructed indicating the response category to which each section of the other vectors corresponds.

We shall look at data, given in Table 3.6, for 21 people with one of three

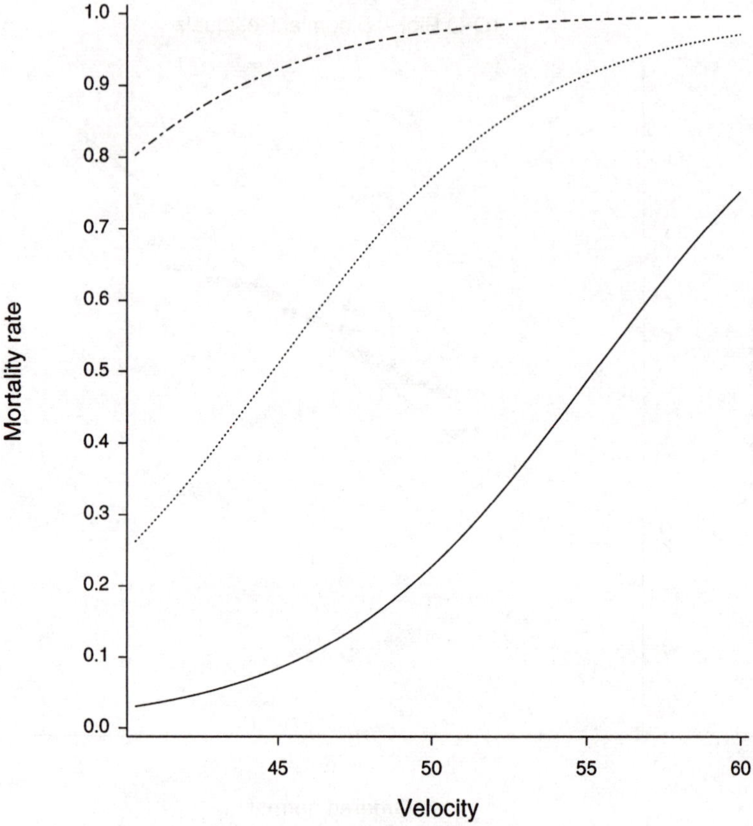

Fig. 3.10. Logistic regression model of Equation (3.6) for velocity at ages 25 (solid), 40 (dots), and 55 (dot–dashes) for the collision data of Table 3.4.

types of Cushing's syndrome, the response. This is a disease associated with the overproduction of cortisone by the adrenal cortex. The explanatory variables are the rates at which two steroid metabolites are excreted in the urine (mg/day).

The vectors for the two metabolites are tripled in length, as `TETRA` and `PREG`. The 'response' variable has 6 ones and then 15 zeros in the first part, 6 zeros, 10 ones, and 5 zeros in the second part, and 16 zeros followed by 5 ones in the third part. The factor variable, `TYPE`, has 21 ones, 21 twos, and 21 threes. To fit the marginal totals, a second factor variable, `EXPL` indexing the 21 different combinations of explanatory variables, is also required.

The model for independence of syndrome type from the metabolite levels,

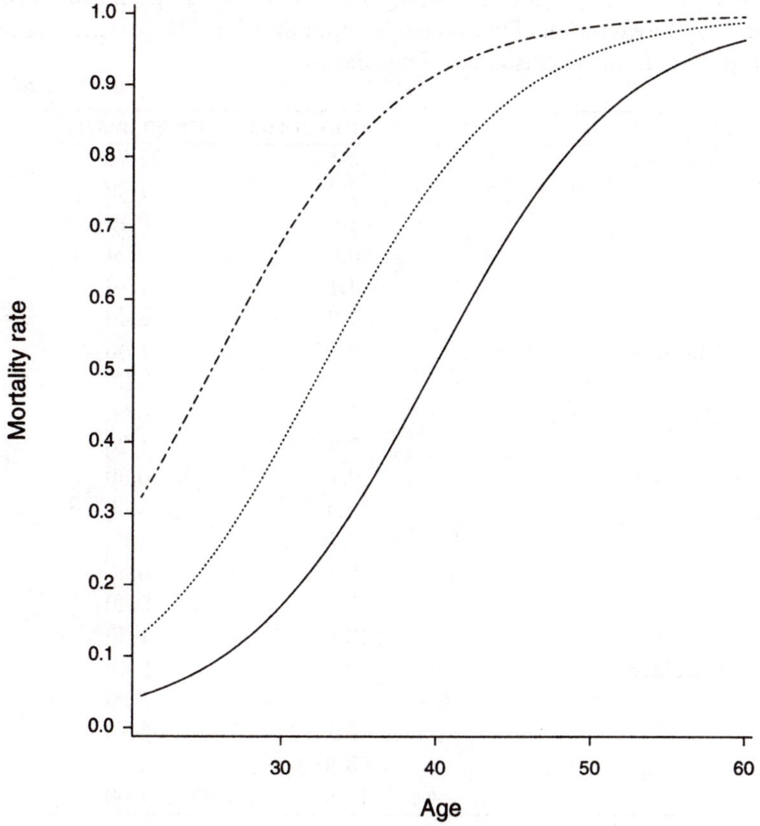

Fig. 3.11. Logistic regression model of Equation (3.6) for age at velocities 45 (solid), 50 (dots), and 55 (dot–dashes) for the collision data of Table 3.4.

$$\text{TYPE} + \text{EXPL}$$

has a deviance of 44.22 (90.22). As with binary data, this deviance has no absolute meaning, and cannot be interpreted. Adding the effect of tetrahydrocortisone

$$\text{TYPE} + \text{EXPL} + \text{TYPE} \cdot \text{TETRA}$$

reduces the deviance by 22.88 (AIC 71.34) with 2 d.f., while that of Pregnanetriol,

$$\text{TYPE} + \text{EXPL} + \text{TYPE} \cdot \text{PREG}$$

reduces it by 5.00 (AIC 87.23) with 2 d.f. Reintroducing the effect of

Table 3.6. Three types of Cushing's syndrome for 21 patients, as they depend on the excretion of two steroid metabolites (mg/day). (Christensen, 1990, p. 271, from Aitchison and Dunsmore)

Type	Tetrahydrocortisone	Pregnanetriol
Adenoma	3.1	11.70
	3.0	1.30
	1.9	0.10
	3.8	0.04
	4.1	1.10
	1.9	0.40
Bilateral hyperplasia	8.3	1.00
	3.8	0.20
	3.9	0.60
	7.8	1.20
	9.1	0.60
	15.4	3.60
	7.7	1.60
	6.5	0.40
	5.7	0.40
	13.6	1.60
Carcinoma	10.2	6.40
	9.2	7.90
	9.6	3.10
	53.8	2.50
	15.8	7.60

Table 3.7. Parameter estimates and standard errors for the final model of Equation (3.7) for the data on Cushing's syndrome in Table 3.6.

Parameter	Estimate	s.e.
TYPE(1)·TETRA	−2.916	2.856
TYPE(2)·TETRA	−0.093	0.090
TYPE(3)·TETRA	0.000	—
TYPE(1)·PREG	0.386	2.512
TYPE(2)·PREG	−1.522	1.099
TYPE(3)·PREG	0.000	—

tetrahydrocortisone,

$$\texttt{TYPE} + \texttt{EXPL} + \texttt{TYPE} \cdot \texttt{TETRA} + \texttt{TYPE} \cdot \texttt{PREG} \tag{3.7}$$

gives a total reduction of 33.65 (AIC 64.57) with 4 d.f. An interaction

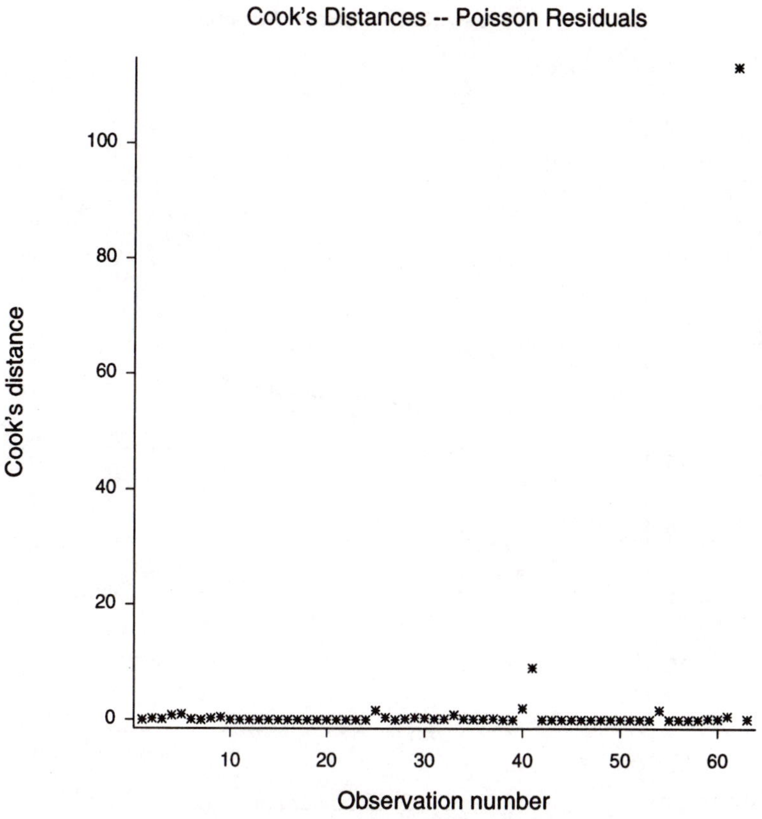

Fig. 3.12. Cook's distances from the polytomous regression model of Equation (3.7) for Cushing's syndrome in Table 3.6.

between the two only changes the deviance by a further 1.32 (AIC 67.24).

It is interesting to note that this is an example where the standard errors of the estimates, given in Table 3.7 for the model of Equation (3.7), are very misleading. None of the parameters appear to be significantly different from zero, and yet the changes in deviance are 26.66 for eliminating `TETRA` and 10.77 for `PREG`. This is not an artifact of the method of fitting; these are the correct standard errors, as were those in Table 3.5 for the binary data example above.

From these estimates, we see that adenoma is associated with low levels of tetrahydrocortisone and carcinoma with high levels, while adenoma has somewhat higher levels of Pregnanetriol, and bilateral hyperplasia much lower levels.

Closer inspection of the data reveals one very high value of tetrahy-

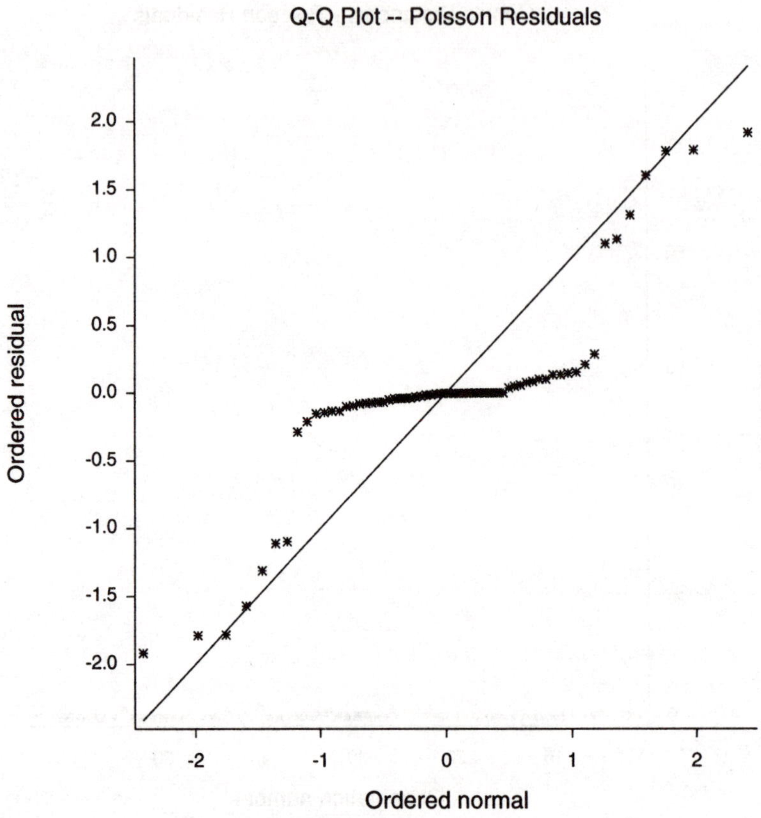

Fig. 3.13. Normal probability plot from the polytomous regression model of Equation (3.7) for Cushing's syndrome in Table 3.6.

drocortisone. This is reflected in the plot of Cook's distances, in Figure 3.12, where this value stands out. Again, this may be an outlier. Here, it would be an error in recording the original data, which should be checked with the original coding sheets and questionnaires. The horizontal middle section of the normal probability plot, in Figure 3.13, simply reflects the large number of zero frequencies in our setup for the polytomous regression. Otherwise, the two ends of this plot appear reasonable.

Because the values of the two explanatory variables are quite skewed, Christensen (1990) suggests taking their logarithms. However, the best model with these transformed variables, the analogue of Equation (3.7), has a deviance of 12.30 (66.30) as compared to 10.57 (64.57) for the model with untransformed variables which we have used.

3.6 Exercises

(1) Can a simpler model be found for the case-control study of oesophageal cancer in Exercise (2.3)?

(2) The table in Exercise (2.5) classified the Inuit by age. See if the models for those data can be simplified by using the regression techniques of this chapter.

(3) Can your model for the coal miner data in Exercise (2.9) be simplified using a different variable for age?

(4) Beetles were exposed to gaseous carbon disulphide and their mortality recorded after five hours (Stukel, 1988, from Bliss):

Dose	Deaths	Total
1.6907	6	59
1.7242	13	60
1.7552	18	62
1.7842	28	56
1.8113	52	63
1.8389	53	59
1.8610	61	62
1.8839	60	80

How does the mortality rate depend on the dose of carbon disulphide?

(5) The proportion of individuals in Bongono, Zaire, with malaria antibodies present was recorded (Morgan, 1992, p. 16) as in the following table:

Mean age	Sero-positive	Total	Mean age	Sero-positive	Total
1.0	2	60	22.0	20	84
2.0	3	63	27.5	19	77
3.0	3	53	32.0	19	58
4.0	3	48	36.8	24	75
5.0	1	31	41.6	7	30
7.3	18	182	49.7	25	62
11.9	14	140	60.8	44	74
17.1	20	138			

Find an appropriate model to describe how this proportion changes with the age of the individuals.

(6) The synergistic effects of two insecticides, called rotenone (mg/l) and deguelin concentrate (mg/l), were tested on chrysanthemum aphis (Vidmar *et al.*, 1992, from Martin):

Rotenone	Deguelin	Deaths	Total
10.2	0.0	44	50
7.7	0.0	42	49
5.1	0.0	24	46
3.8	0.0	16	48
2.6	0.0	6	50
0.0	50.5	48	48
0.0	40.4	47	50
0.0	30.3	47	49
0.0	20.2	34	48
0.0	10.1	18	48
0.0	5.1	16	49
5.1	20.3	48	50
4.0	16.3	43	46
3.0	12.2	38	48
2.0	8.1	27	46
1.0	4.1	22	46
0.5	2.0	7	47

Study the simultaneous influence of the two insecticides on deaths.

(7) The following table gives the proportions of subjects testing positive to toxoplasmosis in 34 cities of El Salvador (Efron, 1986).

Rainfall	Positive	Total	Rainfall	Positive	Total
1735	2	4	1770	33	54
1936	3	10	2240	4	9
2000	1	5	1620	5	18
1973	3	10	1756	2	12
1750	2	2	1650	0	1
1800	3	5	2250	8	11
1750	2	8	1796	41	77
2077	7	19	1890	24	51
1920	3	6	1871	7	16
1800	8	10	2063	46	82
2050	7	24	2100	9	13
1830	0	1	1918	23	43
1650	15	30	1834	53	75
2200	4	22	1780	8	13
2000	0	1	1900	3	10
1770	6	11	1976	1	6
1920	0	1	2292	23	37

How does the possibility of contracting toxoplasmosis depend upon the rainfall where a subject lives?

(8) The age of menarche was determined in a sample of Warsaw girls (Stukel, 1988, from Milicer and Szczotka), as it depended on age:

Age	Menstruating	Total	Age	Menstruating	Total
9.21	0	376	13.33	67	106
10.21	0	200	13.58	81	105
10.58	0	93	13.83	88	117
10.83	2	120	14.08	79	98
11.08	2	90	14.33	90	97
11.33	5	68	14.58	113	120
11.58	10	105	14.83	95	102
11.83	17	111	15.08	117	122
12.06	16	100	15.33	107	111
12.33	29	93	15.58	92	94
12.58	39	100	15.83	112	114
12.83	51	108	17.58	1049	1049
13.08	47	99			

Develop a logistic model to describe these data.

(9) The following table gives the analgesic effect of iontophoretic treatment with vincristine on elderly patients with neuralgia (Piegorsch, 1992). The response variable is whether there was pain or not. Duration is that of the complaint, in months, before treatment began.

Treatment	Age	Sex	Duration	Pain
Y	76	M	36	Y
Y	52	M	22	Y
N	80	F	33	N
Y	77	M	33	N
Y	73	F	17	N
N	82	F	84	N
Y	71	M	24	N
N	78	F	96	N
Y	83	F	61	Y
Y	75	F	60	Y
N	62	M	8	N
N	74	F	35	N
Y	78	F	3	Y
Y	70	F	27	Y
N	72	M	60	N
Y	71	F	8	Y
N	74	F	5	N
N	81	F	26	N

(a) Investigate the influence of treatment on pain.
(b) Do any of the other factors have an effect?

(10) In a study to try to characterize dwellings plagued by noise from the street in Elsinore, Denmark (Andersen, 1991, p. 316), noise level,

sex, number of inhabitants in the dwelling, number of rooms, income (1000 DKr.), and age in years were recorded:

Case	Sex	Inhabitants	Rooms	Income	Age	Noise
1	F	2	3	11	59	None
2	F	4	5	0	53	None
3	F	4	5	16	36	None
4	M	3	5	4	26	Much
5	F	2	2	7	57	None
6	M	3	5	14	51	None
7	M	3	3	8	52	None
8	F	2	3	5	59	None
9	F	2	2	7	31	None
10	F	2	3	9	53	Some
11	M	2	3	2	25	None
12	M	2	3	6	69	None
13	F	2	3	9	63	None
14	F	10	5	13	60	Much
15	M	2	2	20	69	Much
16	F	2	4	13	51	None
17	F	4	4	12	28	None
18	F	2	3	12	41	None
19	M	3	5	8	44	None
20	F	4	7	11	29	Much
21	F	4	3	8	25	None
22	M	4	5	12	32	None
23	F	3	6	10	27	None
24	F	2	4	3	62	None
25	F	2	4	7	42	None
26	F	4	5	20	42	None
27	M	2	5	15	37	None
28	M	3	4	8	28	None
29	M	4	5	13	37	Some
30	F	4	3	14	27	None
31	F	2	4	12	41	Some
32	F	3	5	10	31	Some
33	M	2	3	8	56	None
34	M	4	4	5	40	None
35	M	5	4	6	43	Much
36	M	2	3	14	53	Much

With what factors does noise level vary in these data?

4
Ordinal variables

In the analysis of categorical data, ordinal variables are commonly encountered. The categories are known to have an order but knowledge of the scale is insufficient to consider them as forming a metric. Although they may be treated simply as nominal categories, as in the first two chapters, valuable information is being lost. At the same time, they cannot be handled as continuous variables, using the models of the previous chapter, without making often unjustified assumptions. Occasionally, a linear scale can be used for an ordinal variable, but the goodness of fit of such a model should be checked, as we shall always do below. However, in most cases, special models are required for such variables. Thus, ordinal variables lie in between nominal and continuous variables. In this chapter, we shall employ several approaches to modelling them.

4.1 Extremity models

The simplest models for dependence among variables involve only the symmetry of a very few cells, especially the corner cells. If the variables have an order, as is often the case in such tables, the corners are the extremes, hence the name of this class of models.

Consider a simple two-way table relating opinions on whether one agrees or not that grocery shopping is `TIRING` to availability of a `CAR`, obtained in a survey in Oxford, England, as presented in Table 4.1. We first fit a model for independence between opinions and car availability,

$$\texttt{TIRING} + \texttt{CAR} \tag{4.1}$$

Table 4.1. Oxford shopping survey. (Fingleton, 1984, p. 10)

	Grocery shopping is tiring				
Car available	Disagree	Tend to disagree	In between	Tend to agree	Agree
No	55	11	16	17	100
Sometimes	101	7	18	23	103
Always	91	20	25	16	77

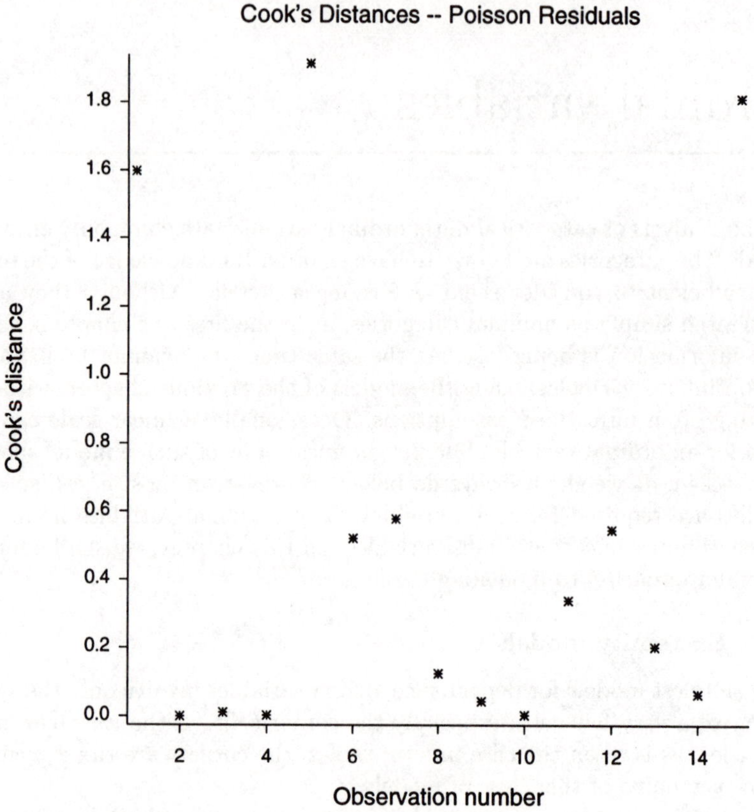

Fig. 4.1. Cook's distances from the independence model of Equation (4.1) for the Oxford shopping data of Table 4.1.

With a deviance of 23.87 (37.87) and 7 d.f., this model is implausible. An alternative would be to fit the saturated model,

$$\texttt{TIRING} + \texttt{CAR} + \texttt{TIRING} \cdot \texttt{CAR}$$

but this uses the maximum number of parameters, and, thus, is too complex to provide much useful information.

From Cook's distances for the independence model, plotted in Figure 4.1, we see that the three observations fitting most poorly are the extremes, those in the two upper and the lower right corners. Let us first look at the two conflictual extremes, disagreeing that shopping is tiring when no car is available and agreeing that shopping is tiring when a car is always available. One possibility is that they are the exceptions to independence, because they should have a lower probability of occurring.

We construct a factor variable with two levels, say CEXT1, contrasting these extremes with the rest of the table:

2	1	1	1	1
1	1	1	1	1
1	1	1	1	2

and add this to the model:

$$\texttt{TIRING} + \texttt{CAR} + \texttt{CEXT1}$$

With a deviance of 10.21 (26.21) and 7 d.f., the model now fits reasonably well. Cook's distances (not shown) no longer indicate a problem with the extremes. The parameter value (−0.401) for this new variable confirms that the two corners have lower probability. Finding grocery shopping tiring does not depend on having a car, except for these two extreme responses, which occur relatively too infrequently.

In principle, we do not need to continue. However, for other data, an additional step might be required. The opposite extreme corners, the concordant ones, might have too high a probability of occurrence. We can then set up a three-level factor variable (CEXT2):

2	1	1	1	3
1	1	1	1	1
3	1	1	1	2

and refit the model to the data:

$$\texttt{TIRING} + \texttt{CAR} + \texttt{CEXT2}$$

For our present example, the model must fit well because the two corner model did; the deviance is now 9.65 (27.65) with 6 d.f. Surprisingly, the parameter value (−0.150) indicates that the two concordant corners also have lower probability than the body of the table, although this time the difference between models is not large.

If a single cell is an extreme case, the easiest way to account for it in a model is by giving it zero weight. This is equivalent to creating a two-level factor variable, where only that one cell has the second factor level.

Thus, we see that, in certain cases, it is possible to localize where dependence is occurring when two or more ordinal variables are related in a table.

4.2 Log multiplicative model I

The first, and perhaps most obvious, way to model an ordinal variable directly is to estimate a scale upon which the values of the variable lie. However, such a scale will never be unique. It must always be calculated in relation to one or more other variables. In this way, the choice of such criterion variables determines the resulting scale, which varies with that choice. The scale is estimated by successive approximations and then, finally, fitted as if it were a continuous variable. In this section, we consider the case of a table with one ordinal variable and one or more nominal variables. In the next section, we apply the same principle to tables with two ordinal variables.

When the scale for a nominal variable is estimated, the model is called log multiplicative, because it is no longer linear in the unknown parameters, but multiplicative on the log scale for these parameters. Two unknown parameters are multiplied:

$$\log(\nu_{ij}) = \mu + \theta_i + \phi_j + \alpha_i \xi \tag{4.2}$$

Both α_i, indexing the nominal variable(s), and ξ, the scale to be estimated, are unknown parameters. Because the model is not linear in these parameters, estimation cannot be done directly with a generalized linear model algorithm, but must proceed iteratively by an extra loop of successive approximations. (A GLIM4 macro for this is supplied in Appendix A.1.)

We shall apply the model to data on criminal cases in North Carolina, given in Table 4.2. The ordinal variable is the outcome of the case and the three explanatory variables are race, type of offence, and county. Although, as it is presented, the case outcome does not appear to be ordinal, we shall see that useful results can be obtained with this approach.

The three explanatory variables will first be combined as one complex variable with 20 categories indexing the rows of the table. This is equivalent to including all possible interactions among these variables, in relation to the ordinal variable, in the model. After we have obtained the scale, we shall verify if all such interactions are necessary, or if some may be eliminated. Thus, our nominal scale will be constructed in relation to this set of three variables.

We first fit the usual model for independence, here between the ordinal variable, outcome of the case, taken as a nominal variable, and the three explanatory variables:

$$\texttt{OUTCOME} + \texttt{OFF} * \texttt{COUNTY} * \texttt{RACE}$$

With a deviance of 156.23 (200.23) and 38 d.f., we see that this independence is decisively rejected. The outcomes of criminal cases in North Carolina depend on one or more of the variables, county, race, and type of offence.

We next treat the variable of interest as if it were metric and linear. This is equivalent to assuming that the ordinal scale has equal intervals between its categories:

$$\texttt{OUTCOME} + \texttt{OFF} * \texttt{COUNTY} * \texttt{RACE} \cdot (1 + \texttt{OUTCOMEL})$$

Table 4.2. Outcomes of criminal cases in North Carolina, classified by type of offence, county, and race. (Upton, 1978, p. 104, from Lehnen and Koch)

			Plea		
Offence	Race	County	Not prosecuted	Guilty	Not guilty
Drinking	Black	Durham	33	8	4
Violence			10	10	3
Property			9	8	2
Traffic			4	2	1
Speeding			32	3	0
Drinking		Orange	5	10	1
Violence			5	5	5
Property			11	5	3
Traffic			12	6	1
Speeding			20	3	2
Drinking	White	Durham	53	2	2
Violence			7	8	1
Property			10	5	2
Traffic			16	3	2
Speeding			87	5	3
Drinking		Orange	14	2	0
Violence			1	5	7
Property			5	4	0
Traffic			13	13	1
Speeding			98	16	7

Here, `OUTCOME` is a factor variable, whereas `OUTCOMEL` is a linear variable.

With a deviance of 40.20 (121.20) and 19 d.f., the fit is improved, showing that outcome is related to race, county, and type of offence, but the lack of fit is still rather large. The ordinal scale appears not to be equally spaced. The slope for speeding offences is smaller than that of the other offences, either negative (−0.515, −0.369) in Durham county for Blacks and Whites, respectively, or about zero (0.0598, 0.000) in Orange county. This offence is prosecuted less often than the other four. Also, the slopes for drinking for Whites are also negative (−0.415, −0.332) in the two counties.

Finally, we fit the log multiplicative model. We see that the estimated scale, (0.000, 0.897, 1.000), places the two prosecution outcomes close together, separated from the no prosecution outcome. Let us call this new estimated scale variable `ORD`. Now, with a deviance of 23.85 (107.85) and 18 d.f., the fit is good, perhaps too good, because a large number of parameters have been included to represent all the interactions. The normal probability plot, given in Figure 4.2, shows a good fit (the extreme residuals are not extreme enough), while Cook's distances, given in Figure 4.3, indi-

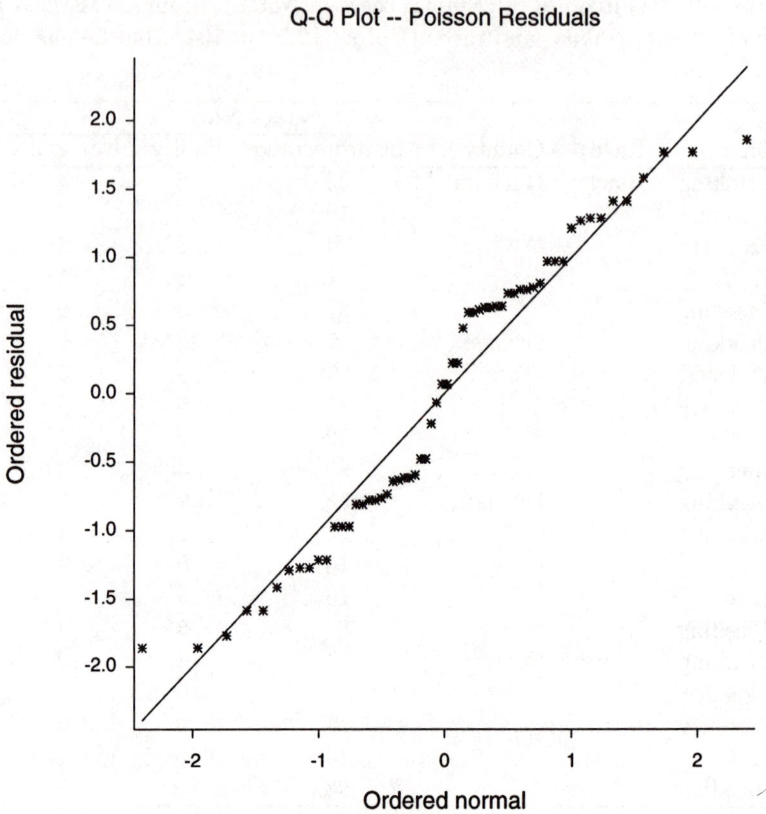

Fig. 4.2. Normal probability plot from the log multiplicative model of Equation (4.2) for the criminal cases data of Table 4.2.

cate a number of categories with large frequencies as fitting poorly. (The three largest are underlined in Table 4.2; they are not the extremes in the normal probability plot.)

The pattern of values for the estimates remains the same. Blacks are proportionately prosecuted less for speeding than for other offences (negative or near zero values: −1.035, 0.083), while Whites are prosecuted less for both drinking (−1.204, −0.580) and speeding (−1.006, 0.000). Orange county prosecutes speeding, independent of race (0.083, 0.000), proportionately more than does Durham (−1.035, −1.006); the slope is flatter in the former as compared to the latter.

Let us now try to simplify our model by fitting only the main effects, i.e. the relation between this ordinal variable and each of the explanatory variables, but none of the interactions among them:

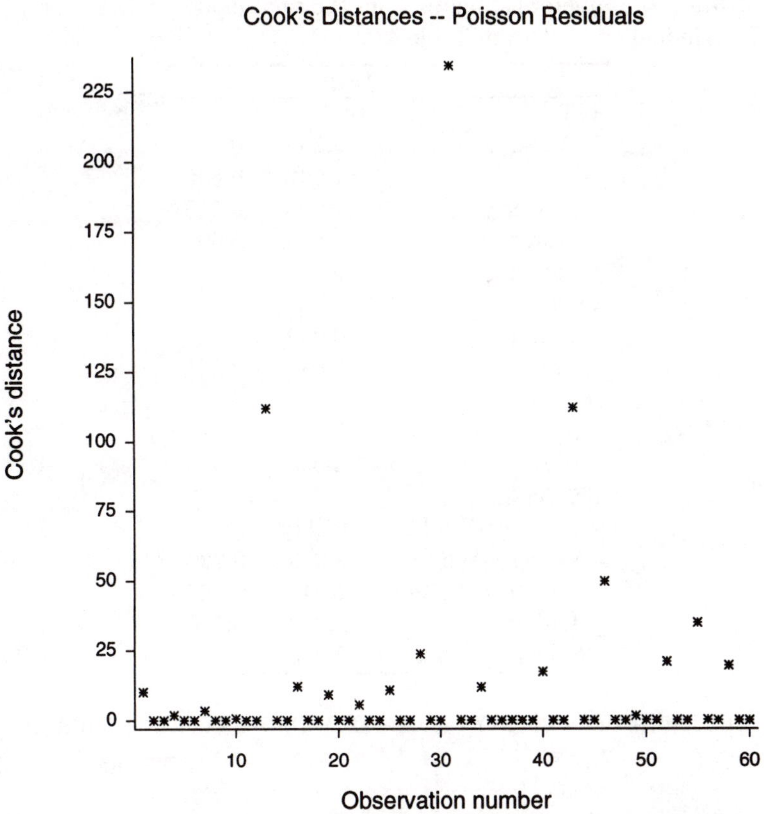

Fig. 4.3. Cook's distances from the log multiplicative model of Equation (4.2) for the criminal cases data of Table 4.2.

$$\mathtt{OUTCOME} + \mathtt{OFF} * \mathtt{COUNTY} * \mathtt{RACE} + \mathtt{ORD} \cdot (\mathtt{OFF} + \mathtt{COUNTY} + \mathtt{RACE})$$

However, we may suspect, from what preceded, that an interaction between type of offence and race will be necessary. With a deviance of 50.90 (108.90) and 31 d.f., the fit is not sufficiently good. We gain 13 d.f., but have eliminated too many parameters. If we look at the parameter estimates, we see that, on average, speeding is prosecuted less often than the other offences, that Orange county prosecutes more, on average, than Durham, and that Whites are prosecuted less often than Blacks.

The different interactions may now be tried; we quickly discover that it is sufficient to add the interaction between type of offence and race in relation to outcome in order to obtain a satisfactory model:

Table 4.3. Parameter estimates from the final model of Equation (4.3) for the criminal cases data of Table 4.2.

	Estimate	s.e.
OFF		
Drinking	1.659	0.511
Violence	2.607	0.555
Property	1.896	0.548
Traffic	1.270	0.608
Speeding	0.000	—
COUNTY		
Durham	0.000	—
Orange	0.882	0.231
RACE		
Black	0.000	—
White	−2.050	0.546
OFF·RACE		
Drinking·White	0.000	—
Violence·White	2.619	0.799
Property·White	1.941	0.787
Traffic·White	2.240	0.776
Speeding·White	2.030	0.718

$$\texttt{OUTCOME} + \texttt{OFF} * \texttt{COUNTY} * \texttt{RACE} + \texttt{ORD} \cdot (\texttt{OFF} + \texttt{COUNTY} + \texttt{RACE} + \texttt{OFF} \cdot \texttt{RACE}) \quad (4.3)$$

with deviance 35.23 (101.23) and 27 d.f. The pertinent parameters of this final model are presented in Table 4.3. Again, we see that offences of violence (2.607) are more often at the high end of the scale (i.e. prosecuted most) and speeding (0.000) at the low end, that Orange County prosecutes more (0.882), and that Whites are prosecuted less (−2.050). The interaction now shows that Whites are prosecuted proportionately less than Blacks for drunkenness; for Whites, the four interaction parameters are all positive as compared to zero for the first category, drinking.

Cook's distances and the residual plots (not shown) are now as expected for an acceptable model: no obvious pattern or very large values in the former, and the residual plot still lying at 45 degrees.

4.3 Log multiplicative model II

When a table contains two ordinal variables, a scale may be estimated for each of them in relation to the other. We still have a log multiplicative model:

$$\log(\nu_{ij}) = \mu + \theta_i + \phi_j + \alpha\xi\omega$$

Table 4.4. Schizophrenic patients in London. (Haberman, 1974b, from Wing)

Years	Goes home or is visited regularly	Visited less than once a month and does not go home	Never visited and never goes home
2–10	43	6	9
10–20	16	11	18
> 20	3	10	16

but now with three unknown parameters multiplied together. The unknown scales are ξ and ω, while α is a regression parameter estimated once the scales are calculated.

Here, we shall take a much simpler table as illustration, the relationship between length of stay for schizophrenic patients in London mental hospitals and frequency of visit, as shown in Table 4.4. Let us first look at the independence model,

$$\texttt{VISIT} + \texttt{LENGTH}$$

which has a deviance of 38.35 (48.35) with 4 d.f. This model must be rejected. Frequency of visit depends on years of internment; the question is, how?

A second possible model takes both variables as metric and linear, i.e. as equally spaced scales:

$$\texttt{VISIT} + \texttt{LENGTH} + \texttt{VISITL} \cdot \texttt{LENGTHL}$$

It has a deviance of 7.12 (19.12) with 3 d.f. Again, the fit is improved, but is not really acceptable (the AIC for the saturated model is 18). The parameter estimate, 0.197, is the slope, which, being positive, shows that the two variables vary together. However, frequency of visit decreases from left to right in the table, so that it decreases with increasing length of stay.

The following two models will fit, the first successfully, one variable as metric and linear and the other as nominal, then the reverse. When length of stay is linear,

$$\texttt{VISIT} + \texttt{LENGTH} + \texttt{VISIT} \cdot \texttt{LENGTHL}$$

the fit is quite acceptable, with a deviance of 0.02 (14.02), while it is not acceptable in the second model,

$$\texttt{VISIT} + \texttt{LENGTH} + \texttt{VISITL} \cdot \texttt{LENGTH}$$

with a deviance of 6.46 (20.46) and 2 d.f., where frequency of visit has the

Table 4.5. Deviance table comparing the models fitted to the schizophrenic patients data of Table 4.4.

Effect	Deviance	d.f.
Linear effects	31.23	1
Visits	7.10	1
Stay	0.66	1
Other effects(1)	0.00	1
Other effects(2)	0.01	1

equal interval scale. In the first of these models, we see that longer length of stay is less probable for the first category of visits (−0.815) as compared to the other two categories with more or less zero slopes.

A row and column effect model combines the previous two models

$$\mathtt{VISIT} + \mathtt{LENGTH} + \mathtt{VISIT} \cdot \mathtt{LENGTHL} + \mathtt{VISITL} \cdot \mathtt{LENGTH}$$

and, of course, fits well because the first of these did. It has a deviance of 0.002 (16.00) with 1 d.f. However, it is not acceptable, because it is more complex than that with only length of stay on a linear scale.

Finally, the model with the two ordinal variable scales estimated can be fitted. The deviance is 0.01 (16.01) with 1 d.f. With parameter estimates (0.000, 0.512, 1.000) for the scale for length of stay, the model shows that, as already concluded, this variable has virtually equal intervals, while we now see clearly what we suspected from the model with only length of stay being linear: the scale (0.000, 0.984, 1.000) for frequency of visit contrasts 'goes home or visited regularly' with the two cases of few or no visits and not going home.

These results can be summarized in the analysis of association table which partitions the deviances, as shown in Table 4.5. The 'linear effects' refer to the linear dependence between the two variables, while the 'visits effect' refers to the lack of linearity of the frequency of visit variable. Both are likely to be different from zero, as we have seen. The 'stay' effect looks at linearity of length of stay. The two 'other effects' are interchangeable and only one should be interpreted. They concern any lack of fit remaining when both visit and stay effects are included in the model.

We shall now study the same table with two other models for ordinal variables.

4.4 Proportional odds model

Because the categories of an ordinal variable are, by definition, ordered, frequencies of response in succeeding categories may be compared. This and the following section present two commonly used approaches to such comparisons when there is one ordered response variable.

Table 4.6. The reconstructed table for the proportional odds model, from the schizophrenic patient data of Table 4.4, where (a), (b), and (c) refer to the three columns of that table.

	Goes home more	Goes home less
	(a)	(b)+(c)
2–10	43	15
10–20	16	29
> 20	3	26
	(a)+(b)	(c)
2–10	49	9
10–20	27	18
> 20	13	16

In the proportional odds model, we consider each category in turn and compare the frequency of response at least up to that point on the ordinal scale to the frequency for all points higher on the scale. The first category is compared to all the rest combined, then the first and second combined are compared to all the rest combined, and so on. In this way, the original table with an I category ordinal scale is converted into a series of $I - 1$ subtables, each with a binary categorization, lower/higher than the (sliding) point on the scale. We then have three types of variable, the new binary response variable, indicating more or less on the ordinal scale, a variable indexing the subtables, corresponding to the points on the ordinal scale, and the explanatory variables. An advantage of this construction is that the interpretation of conclusions is not modified when the number of ordinal response categories is changed.

It might appear from the construction of this table that we now have a simple case where the logistic model could be applied to the binary response variable. However, if the observations in the original table were independent, the categories in the new reconstructed table will no longer be. A more complex analysis is called for, one which does not fall into the standard generalized linear model framework. To define the necessary model, what is called a composite link function is required. (A GLIM4 macro is given in Appendix A.2.)

We shall now apply the proportional odds model to our data on schizophrenic patients of the previous section. The reconstructed data are given in Table 4.6. Because we have already seen that length of stay may be treated as an equal interval (linear metric) scale, we use that here. With a deviance of 6.69 (12.69) and 6 d.f., the model fits well, although, according to the AIC, not as well as the log multiplicative model above. The negative value, -1.243, for the parameter estimate indicates that the odds of receiving more rather than less visits decreases with increasing length of stay, confirming the previous results.

Table 4.7. The reconstructed table for the continuation ratio model, from the schizophrenic patient data of Table 4.4, where (a), (b), and (c) refer to the three columns of that table.

	Goes home more	Goes home less
	(a)	(b)
2–10	43	6
10–20	16	11
> 20	3	10
	(a)+(b)	(c)
2–10	49	9
10–20	27	18
> 20	13	16

4.5 Continuation ratio model

The continuation ratio model resembles the proportional odds model closely. A series of subtables is also constructed here. But now, for each category of the ordinal variable considered in turn, the frequency of response, at least up to that point on the ordinal scale, is compared only to the frequency for the immediately following category. The first category is compared to the second, the first and second combined to the third, and so on. Given that the response is at least at some level, what is the chance of continuing to the level immediately following? Again, the original table with an I category ordinal scale is converted into a series of $I - 1$ subtables.

In contrast to the proportional odds model, with the continuation ratio model, independence among observations is retained when the table is reconstructed, and the logistic model may be directly applied. We simply reconstruct the table and fit this model,

$$\texttt{LENGTH} + \texttt{SUBTABLE}$$

using the binomial distribution. For our data on schizophrenic patients, this reconstruction is shown in Table 4.7. When applied to these data, the model also fits well. The deviance is 1.97 (9.97) with 2 d.f., with length of stay as a nominal factor variable. With the equal interval scale for length of stay,

$$\texttt{LENGTHL} + \texttt{SUBTABLE}$$

the result is very similar to that for the previous section, with the same kind of interpretation. The deviance is 2.69 (8.69) with 3 d.f. and the parameter value is −2.357. This model gives the best fit of the three to these data.

The choice among the three models, the log multiplicative, the proportional odds, and the continuation ratio, in the analysis of ordinal variables presented in this chapter is rarely obvious. As seen here, the three are often mutually reinforcing, and not all would be necessary in most situations. The log multiplicative model is often attractive because it provides a scale for one or more ordinal variables. On the other hand, with the proportional odds model, interpretation of results will not be modified if categories are combined but no standard software is presently available. The continuation ratio model is perhaps most specialized, being applicable where one is interested in continuation to each successively higher point on a scale, but it is the most easily fitted in terms of programming time; it only uses a standard logistic model. Greenland (1994) provides further discussion of the various domains of application of these models.

4.6 Exercises

(1) Patients undergoing chemotherapy were categorized by the severity of nausea and by whether or not they received cisplatinum (Farewell, 1982):

	None	Mild	Moderate			Severe
Cisplatinum	0	1	2	3	4	5
No	43	39	13	22	15	29
Yes	7	7	3	12	15	14

Is there any difference between the two treatments?

(2) The table below gives the attitude towards abortion and schooling of a sample of people (Agresti, 1984, p. 157).

	Attitude		
Education	Disapprove	Middle	Approve
< High school	209	101	237
High school	151	126	426
> High school	16	21	138

Find a simple model to describe these data.

(3) The following table shows the mental health status and parental socio-economic status (SES) of a sample of young residents of midtown Manhattan, U.S.A., living at home (Haberman, 1979, p. 375, from Srole *et al.*).

	Mental health category			
SES	Well	Mild	Moderate	Impaired
A	64	94	58	46
B	57	94	54	40
C	57	105	65	60
D	72	141	77	94
E	36	97	54	78
F	21	71	54	71

How does mental health status vary with SES?

(4) The table below shows the severity of pneumoconiosis as related to the number of years working at the coal face (Ashford, 1959).

	Pneumoconiosis		
Years	Normal	Mild	Severe
0.5–11	98	0	0
12–18	51	2	1
19–24	34	6	3
25–30	35	5	8
31–36	32	10	9
37–42	23	7	8
43–49	12	6	10
50–53	4	2	5

Does the severity of pneumoconiosis increase with the number of years worked?

(5) In a clinical trial on the treatment of small cell lung cancer, tumour response in patients receiving chemotherapy was studied under two therapies (Holtbrügge, and Schumacher, 1991). In the sequential group, the same combination of chemotherapeutic agents was administered in each treatment cycle, while in the alternating group, three different combinations were given, alternating from cycle to cycle.

Sex	Therapy	Progressive disease	No change	Partial remission	Complete remission
Male	Sequential	28	45	29	26
Female		4	12	5	2
Male	Alternating	41	44	20	20
Female		12	7	3	1

(a) Which treatment shows better results?

(b) Is there any evidence of difference between the sexes?

(6) Study the relationship of drinking habits of subjects living in group quarters to the location in New York, U.S.A., and the length of time they have been there (Upton, 1978, p. 104).

Number of years in quarters	Location	Light	Moderate	Heavy
0		25	21	26
1–4	Bowery	21	18	23
5+		20	19	21
0		29	27	38
1–4	Camp	16	13	24
5+		8	11	30
0		44	19	9
1–4	Park Slope	18	9	4
5+		6	8	3

(7) An experiment was performed to study the red core disease in strawberries, caused by the fungus, *Phytophtora fragariae* (Jansen, 1990). Twelve groups were obtained by crossing three genotypes (male parent) with four other genotypes (female parent). The numbers of strawberry plants with damage caused by the fungus were classified into three ordered categories, as given in the following table:

		Block											
		1			2			3			4		
Genotype		Disease severity											
Male	Female	1	2	3	1	2	3	1	2	3	1	2	3
1	1	0	3	6	2	2	6	2	3	5	2	5	3
1	2	2	3	5	0	3	7	4	6	0	2	3	5
1	3	3	4	3	7	2	1	1	1	7	2	3	5
1	4	0	5	5	5	4	1	2	8	0	1	4	5
2	1	1	4	4	2	2	6	1	2	7	1	5	4
2	2	1	4	5	3	4	2	1	6	3	4	2	4
2	3	4	3	3	5	1	4	3	3	4	4	2	4
2	4	1	4	5	1	2	6	8	2	0	2	5	3
3	1	0	0	9	3	5	2	2	5	3	0	0	10
3	2	5	3	2	3	2	5	3	6	1	2	1	7
3	3	0	3	6	2	5	3	1	3	6	0	3	7
3	4	3	0	7	5	2	3	7	3	0	3	4	3

Determine if the incidence of disease varies among the groups.

(8) A randomized, double bind clinical trial was conducted to compare an active hypnotic drug with placebo in patients with insomnia. Patients were asked how long it took them to fall asleep, both after a one week placebo washout period (baseline) and after a two week treatment period (Francom *et al.*, 1989):

	Treatment							
	Active				Placebo			
	After Treatment							
Base	< 20	< 30	< 60	> 60	< 20	< 30	< 60	> 60
< 20	7	4	1	0	7	4	2	1
< 30	11	5	2	2	14	5	1	0
< 60	13	23	3	1	6	9	18	2
> 60	9	17	13	8	4	11	14	22

(a) Does the hypnotic drug have a superior effect to the placebo?

(b) Does any effect observed depend on the initial state of the patient?

(9) Dumping severity, an undesirable sequel of surgery for duodenal ulcer, was classified by the type of operation for four hospitals (Grizzle *et al.*, 1969):

		Dumping severity		
Operation	Hospital	None	Slight	Mod.
Drainage and vagotomy	1	23	7	2
25% resection and vagotomy		23	10	5
50% resection and vagotomy		20	13	5
75% resection		24	10	6
Drainage and vagotomy	2	18	6	1
25% resection and vagotomy		18	6	2
50% resection and vagotomy		13	13	2
75% resection		9	15	2
Drainage and vagotomy	3	8	6	3
25% resection and vagotomy		12	4	4
50% resection and vagotomy		11	6	2
75% resection		7	7	4
Drainage and vagotomy	4	12	9	1
25% resection and vagotomy		15	3	2
50% resection and vagotomy		14	8	3
75% resection		13	6	4

(a) Study how dumping severity depends on the type of surgery.

(b) Check if it varies among hospitals.

(10) In Exercise (3.10), the noise level is an ordinal variable. Can a model be developed to take this into account?

5
Zero frequencies

When several categorical variables are cross-tabulated to form a table of several dimensions, some cells often contain zero frequencies of response. Such a situation may arise in at least three ways. The zero may have occurred simply because the sample size was not large enough and that combination of categories is not represented. A large enough sample would theoretically include such combinations. In the other two cases, the combination of categories is actually impossible, either a structural zero or a model zero, the latter being produced by excluding the combination from the model for some theoretical reason. In the first case, the expected frequencies or fitted values should be positive. In the latter two cases, the expected frequencies for any model must be zero. In the next section, we shall consider the first case, that of sampling zeros. In subsequent sections, as well as in Part II, various forms of incomplete tables with structural and model zeros will be covered.

5.1 Sampling zeros

If a saturated model is to be fitted, even one sampling zero will create problems for estimation of the parameters. Most software will only estimate as many parameters as there are non-zero entries in the table. For example, if a sweep algorithm is used for estimation, the last parameters encountered in the list will not be estimated, even though this may have no actual relationship to the location of the zeros in the table.

For an unsaturated model, all of the parameters may often be estimated, as long as the number of non-zero frequencies is at least as large as the number of parameters to be estimated. However, unexpected exceptions to this may occur, depending on the location of the zeros in the table. The user must beware, especially if the software has also tabulated the table so that it has not been inspected beforehand. One reliable indication that a problem has occurred is if certain standard errors of parameter estimates are greatly inflated. This does not, however, indicate which zeros are causing the problem. A solution is to try to eliminate those zero cells for which the estimated frequencies are either very large or very small. (Remember that, in the situation of sampling zeros, the estimated frequencies should be

Table 5.1. Survival of melanoma patients, as related to sex, family history of melanoma, and remission status. (adapted from Lee, 1992, pp. 56–59)

		Remission status					
		Relapsed		Still in remission		Never	
	Family	Survival					
Sex	history	Alive	Dead	Alive	Dead	Alive	Dead
Male	No	10	4	0	3	2	5
	Yes	0	0	0	0	0	0
	Unknown	4	0	1	3	6	16
Female	No	7	0	2	4	5	4
	Yes	1	0	1	0	0	1
	Unknown	6	0	2	1	6	5

positive.) These cells are removed by giving them zero weight. Note that the cells to be eliminated will depend on the model fitted to the data and not simply on the location of the zeros in the table.

In removing cells from the table in this way, the degrees of freedom, as indicated by the software, will be reduced by the number of cells eliminated. This should not be taken to mean that this is the correct number of degrees of freedom. There is some debate on this subject, because such a reduction would mean that the degrees of freedom were random.

A simple example illustrates the problem. Table 5.1 gives survival of patients with melanoma as it depends on sex, remission status, and family history of the disease. Note the large number of zeros in this table, including a whole line, that for males with a family history of the disease.

The model for independence of survival from the three explanatory variables,

$$\texttt{SURV} + \texttt{SEX} * \texttt{HIST} * \texttt{REMIS} \tag{5.1}$$

has a deviance of 42.36 (80.36) with 17 d.f. However, if we inspect the parameter estimates and standard errors, given in Table 5.2, we find a number of very large standard errors, all of those where `HIST(2)` is present. (These values are from GLIM4; other software may give different results.) If we weight out the combination, males with a family history of melanoma, which is the row of zeros in the table, we still obtain the same deviance (although the indicated degrees of freedom are smaller). The new parameter estimates and standard errors are also given in Table 5.2. We see that the latter are no longer inflated, that certain parameters are no longer estimated, that the others which formerly had inflated standard errors have changed in value, and that the remaining estimates and standard errors

Table 5.2. Parameter estimates from the independence model of Equation (5.1) for the melanoma data in Table 5.1.

	Without weight		With weight	
	Estimate	s.e.	Estimate	s.e.
1	2.014	0.28	2.014	0.28
SURV(2)	−0.142	0.20	−0.142	0.20
SEX(2)	−0.693	0.46	−0.693	0.46
HIST(2)	−10.64	33.10	−1.946	1.07
HIST(3)	−1.253	0.57	−1.253	0.57
REMIS(2)	−1.540	0.64	−1.540	0.64
REMIS(3)	−0.693	0.46	−0.693	0.46
SEX(2)·HIST(2)	8.692	33.11	0.000	—
SEX(2)·HIST(3)	1.099	0.79	1.099	0.79
SEX(2)·REMIS(2)	1.386	0.85	1.386	0.85
SEX(2)·REMIS(3)	0.945	0.68	0.945	0.68
HIST(2)·REMIS(2)	1.540	46.81	0.154	1.52
HIST(2)·REMIS(3)	0.693	46.81	−0.251	1.50
HIST(3)·REMIS(2)	1.540	0.95	1.540	0.95
HIST(3)·REMIS(3)	2.398	0.71	2.398	0.71
SEX(2)·HIST(2)·REMIS(2)	−1.386	46.83	0.000	—
SEX(2)·HIST(2)·REMIS(3)	−0.945	46.83	0.000	—
SEX(2)·HIST(3)·REMIS(2)	−2.079	1.31	−2.079	1.31
SEX(2)·HIST(3)·REMIS(3)	−2.043	1.01	−2.043	1.01

are unchanged.

If we now go on to fit dependence models, we discover that survival depends on sex,

$$\texttt{SURV} + \texttt{SEX} * \texttt{HIST} * \texttt{REMIS} + \texttt{SURV} \cdot \texttt{SEX}$$

with a reduction in deviance of 5.79 (AIC 70.57) and 1 d.f., and on remission status,

$$\texttt{SURV} + \texttt{SEX} * \texttt{HIST} * \texttt{REMIS} + \texttt{SURV} \cdot \texttt{SEX} + \texttt{SURV} \cdot \texttt{REMIS} \tag{5.2}$$

with a further reduction of 26.96 and 2 d.f. Survival does not depend on family history. The final deviance is 9.61 (47.61) with 14 d.f. (11 d.f. shown by the software). However, this has little absolute meaning because of the sparseness of the table. The plot of Cook's distances, in Figure 5.1, indicates that the final model does not fit very well for the very first observation in the table. The normal probability plot, in Figure 5.2, lies at less than 45°, so that the model may be somewhat overfitted, i.e. have too many parameters. Indeed, if the second and third categories of remission status are amalgamated as REMIS2, so that the relapsed group is contrasted

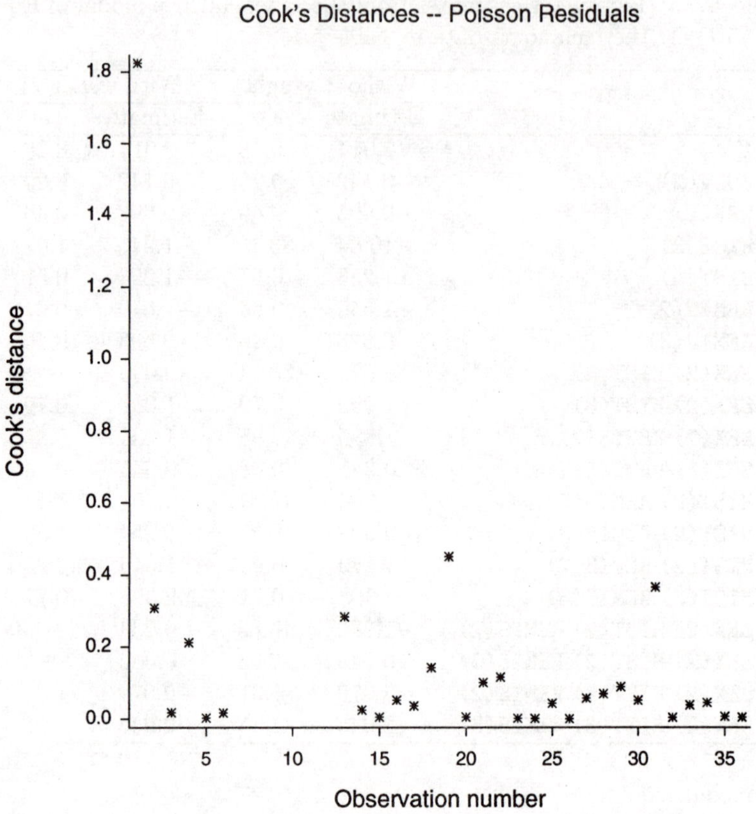

Fig. 5.1. Cook's distances from the model of Equation (5.2) for the melanoma data of Table 5.1.

with the other two,

$$\mathtt{SURV} + \mathtt{SEX} * \mathtt{HIST} * \mathtt{REMIS} + \mathtt{SURV} \cdot \mathtt{SEX} + \mathtt{SURV} \cdot \mathtt{REMIS2}$$

the deviance is only increased by 0.38 (AIC 45.99). Males have less chance of surviving (−1.373), as do those who have relapsed (−2.722).

In this example, the row of zeros eliminated for the independence model allowed the other models of interest to be fitted, even though a considerable number of other zeros was present in the table. In other cases, further zeros may need to be removed as more complex models are fitted.

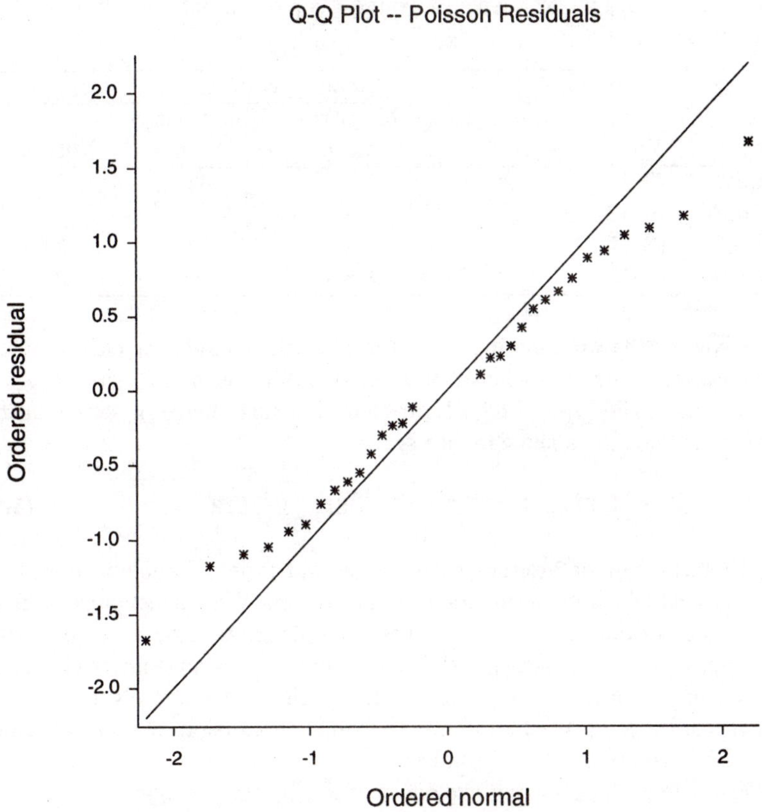

Fig. 5.2. Normal probability plot from the model of Equation (5.2) for the melanoma data of Table 5.1.

5.2 Quasi-independence

When a table involves structural zeros, these are simply not included in the data and the software will fit the model without problem. Our example, given in Table 5.3, involves health problems, as related to the sex and age of young people. The combination male with menstruation problems is impossible, yielding two structural zeros.

We shall use the conventional constraints for these data, i.e. with parameters summing to zero. The minimal (independence) model has the three sets of mean parameters and the sex·age interaction between the two explanatory variables:

$$\mathtt{HEALTH} + \mathtt{SEX} * \mathtt{AGE}$$

Table 5.3. Health problems of young people. (Fienberg, 1977, p. 116, from Brunswick)

		Health problem			
Sex	Age	Sex and reproduction	Menstru- ation	How healthy I am	Nothing
Male	12–15	4	—	42	57
Male	16–17	2	—	7	20
Female	12–15	9	4	19	71
Female	16–17	7	8	10	31

When some cells are missing in a table, such a model is called quasi-independent. With a deviance of 22.02 (36.02) and 4 d.f., the model is to be rejected; the type of health problem depends on age or sex or both.

Next we introduce the effect of sex:

$$\texttt{HEALTH} + \texttt{SEX} * \texttt{AGE} + \texttt{SEX} \cdot \texttt{HEALTH} \tag{5.3}$$

Note that sex cannot interact with the second type of problem, menstruation, so this will not be included in the model. With a deviance of 9.43 (27.43) and 5 d.f., the model now fits satisfactorily, being a considerable improvement over complete quasi-independence. Boys have fewer problems with sex and reproduction (−0.367) than girls, and relatively more with general health (0.385), while about the same proportion of each sex state that they have no problem (−0.018).

We continue by replacing the effect of sex by that of age:

$$\texttt{HEALTH} + \texttt{SEX} * \texttt{AGE} + \texttt{AGE} \cdot \texttt{HEALTH}$$

With a deviance of 13.45 (33.45) and 4 d.f., this model does not fit sufficiently well and must be rejected. Apparently, the type of health problem does not depend on age, at least for the two age groups considered here. It does not seem necessary to continue. Our second model, containing sex, but not age, influencing health problems, will be retained.

Let us, however, look at the residual plots. That for Cook's distances, in Figure 5.3, shows that a fair number of frequencies do not fit well. And, the normal probability plot, in Figure 5.4, does not follow the 45° line, also indicating some lack of fit. If we put dependence on both variables into the model, we have a deviance of 2.03 (26.03), with 2 d.f. This gives a difference of 7.40 and 3 d.f. with respect to the model containing only sex. Although including age yields only a marginal improvement, as indicated by the deviance, it should probably be included because of the poor residual plots without it.

Note that, if the interaction between sex and age with respect to health

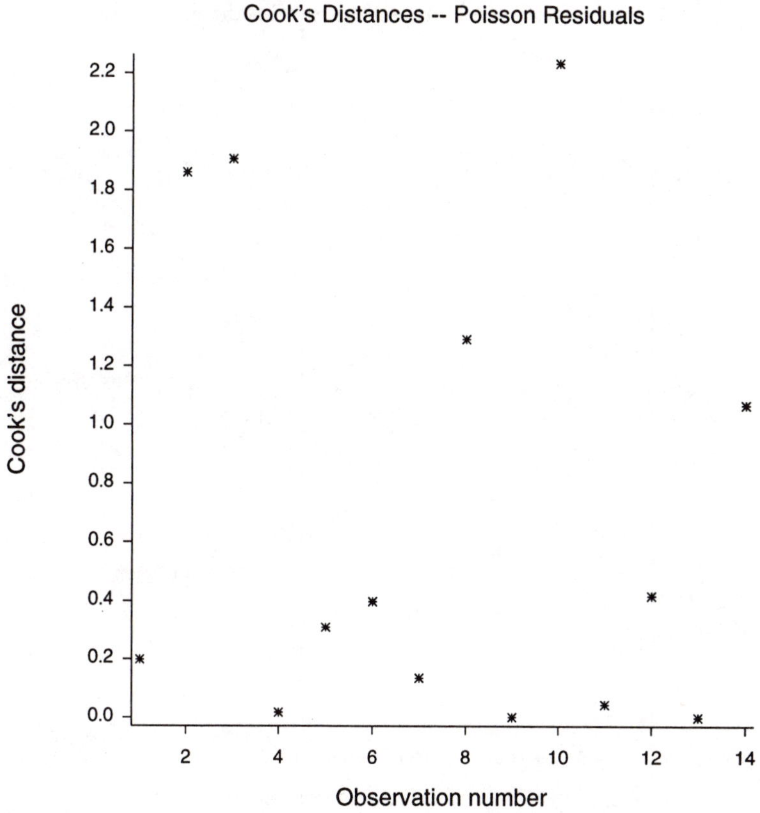

Fig. 5.3. Cook's distances from the model of Equation (5.3) for the health problem data of Table 5.3.

problems were to be included, it would also contain only two parameters, as for sex with problems, because, again, the second health problem does not interact with sex.

5.3 Quasi-symmetry

One important special case of quasi-independence occurs in square tables where the diagonal values are missing. Such tables can be produced when two similar variables are observed, perhaps under similar circumstances. Consider the following example. Each of a group of 37 subjects in a psychology experiment was given a series of 18 messages, three from each of six different categories. They were to return a message, most unlike that received, from a deck containing messages in the five categories not received. Thus, the received and returned messages could not be in the same class.

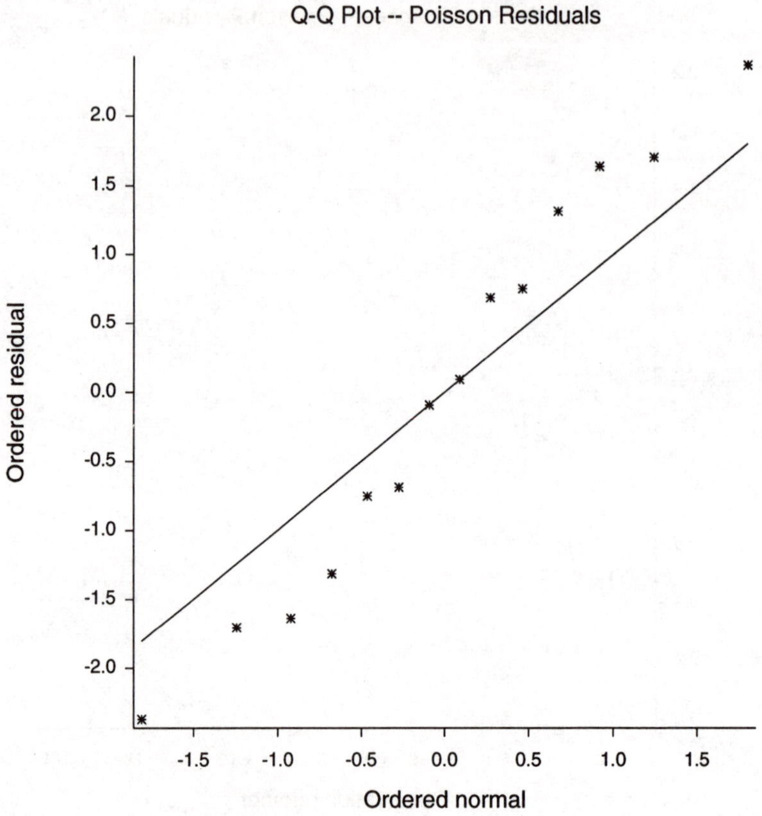

Fig. 5.4. Normal probability plot from the model of Equation (5.3) for the health problem data of Table 5.3.

Such data are presented in Table 5.4. Note that the sum of each row is fixed at $(3 \times 37 =)$ 111.

For this table, the quasi-independence model

$$\texttt{RECEIVE} + \texttt{RETURN}$$

has a deviance of 183.55 (205.55) with 19 d.f., indicating a very poor fit.

Because of the similarity between the two variables in the table, a model can be developed using the symmetry about the main diagonal. We construct a factor variable, `SYM`, of the following form:

Table 5.4. Most unlike resource returned for resource received. (Bishop *et al.*, 1975, p. 299, from Foa)

	Returned					
Received	Love	Status	Information	Money	Goods	Services
Love	—	5	21	48	29	8
Status	4	—	19	27	30	31
Inform.	20	11	—	20	25	35
Money	56	10	21	—	4	20
Goods	42	18	27	6	—	18
Services	12	20	37	26	16	—

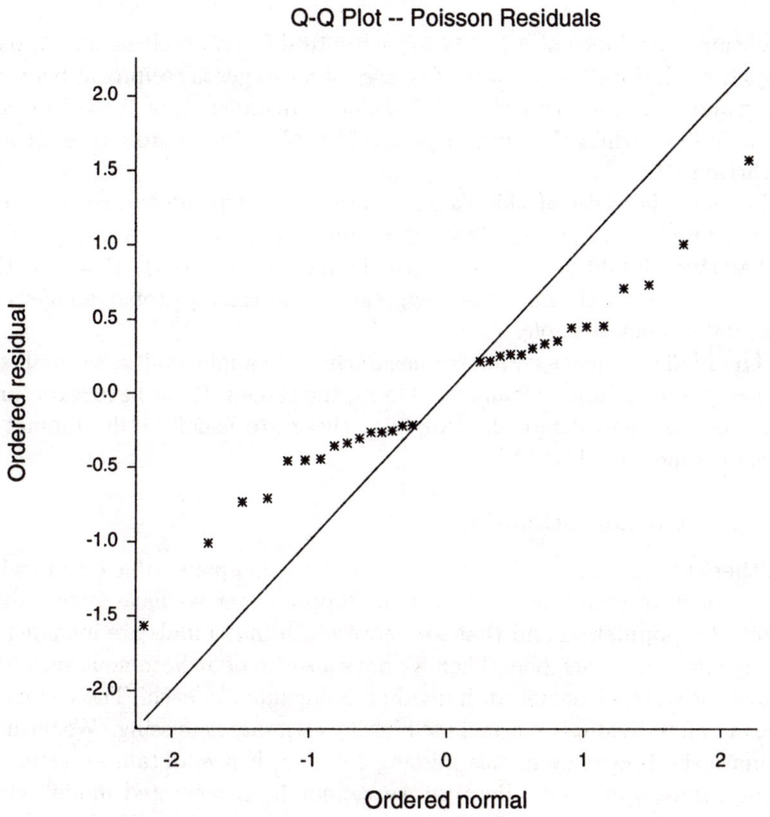

Fig. 5.5. Normal probability plot from the quasi-symmetry model of Equation (5.4) for the resource receiving data of Table 5.4.

—	1	2	3	4	5
1	—	6	7	8	9
2	6	—	10	11	12
3	7	10	—	13	14
4	8	11	13	—	15
5	9	12	14	15	—

The idea is that, when a message of class i is received, the probability of returning a message of class j will be the same as if a message of class i is returned when one of class j is received. However, this cannot be exactly so in the present situation, because the row totals are fixed by the study design.

When we fit the resulting quasi-symmetry model,

$$\texttt{RECEIVE} + \texttt{RETURN} + \texttt{SYM} \tag{5.4}$$

we obtain a deviance of 4.17 (44.17) with 10 d.f., an excellent fit. Apparently, the relationship for pairs of classes of messages is reciprocal between giving and receiving them. The plot of Cook's distances (not shown) reveals no anomalies, while the normal probability plot, in Figure 5.5, indicates overfitting.

In fact, the data of this example are not independent because each subject provides 18 events. Overdispersion (Chapter 10) may be present, so that they should be analysed with the models of Part II. However, the events of all subjects have been aggregated, so that a proper analysis of these data is not possible.

The model of quasi-symmetry has fairly wide applicability; we shall see further uses in Chapters 9 and 10. Often, the table will not have structural zeros on the main diagonal. However, these are usually still eliminated when this model is fitted.

5.4 Population estimation

Another special case of structural zeros in an incomplete table occurs with the problem of population estimation. Suppose that we have several estimates of a population and that we know which individuals are included in the calculation of each one. Then we have a series of dichotomous variables indicating whether or not an individual is included in each. The cell categorizing all individuals not included in any estimate is missing. We wish to estimate the frequency in this missing cell and, hence, obtain an estimate of the total population. Because we cannot fit a saturated model when one category is missing, we must make some assumption of independence among the different types of population estimates used. Under high order independence,

Table 5.5. Data for the estimation of the number of formal volunteer organizations. (Bishop *et al.*, 1975, p. 243)

		Census	
Newspaper	Telephone	Yes	No
Yes	Yes	4	1
No	Yes	8	2
Yes	No	16	49
No	No	113	—

$$\frac{\nu_{111}\nu_{122}\nu_{212}\nu_{221}}{\nu_{222}\nu_{112}\nu_{121}\nu_{211}} = 1$$

for example, with three population estimates. Then, if ν_{222} is the frequency of the missing cell,

$$\nu_{222} = \frac{\nu_{111}\nu_{122}\nu_{212}\nu_{221}}{\nu_{122}\nu_{121}\nu_{211}}$$

Because we shall usually require an interval of plausible values for the total population estimate, N, we need an estimate of the variance (or standard deviation), which can be calculated from

$$\text{var}(N) = \frac{n_{\bullet\bullet\bullet}\nu_{222}}{\nu_{112} + \nu_{121} + \nu_{211} + \nu_{111}} \qquad (5.5)$$

These quantities are easily obtained with most generalized linear modelling software. The missing frequency will be given directly as a fitted value, if an arbitrary value is supplied and given zero weight for the fit. The variance may then be calculated from formula (5.5).

As an example, we take the estimation of the number of formal volunteer organizations in towns in Massachusetts, U.S.A., given in Table 5.5. The three sources of estimates are newspapers, telephones, and a census.

The model with no three-factor interaction,

CENSUS * NEWS + CENSUS * PHONE + NEWS * PHONE

fits perfectly and we easily discover that the interaction between census and newspapers may be eliminated. We fit this model,

CENSUS * PHONE + NEWS * PHONE

which still has a deviance of zero (AIC 12), and calculate the population estimation and its standard deviation. From the fitted values, we see that the number of missing organizations is estimated as 346, so that the total number is estimated as 539. A reasonable interval covers two standard deviations, giving (379, 699) for the total number of organizations. If we also eliminate the newspaper·telephone interaction, the model

$$\texttt{CENSUS} * \texttt{PHONE} + \texttt{NEWS}$$

might still be acceptable, with a deviance of 3.82 (13.82) and 2 d.f., although the AIC indicates a poorer model. One important result of simplifying the model is that the standard deviation is always smaller (Bishop *et al.*, 1975, p. 242). Of course, we must still retain an acceptable model. With the present model, the missing number is 287, so that the estimated total is now 480 and our interval becomes (347, 611), which is not completely within the previous interval. Further simplification of the model, by eliminating the census·telephone interaction, is not possible.

The parameter estimates indicate that there is a positive association between being covered by the census and being listed in the telephone directory (2.344), while coverage by newspapers is relatively independent of both of these.

5.5 Guttman scale model

A Guttman scale is constructed from a series of ordered yes/no questions such that once an individual replies yes (or no) to one question in the series, he/she should also reply yes (or no) to all subsequent questions. A typical series of questions to measure attachment to one's neighbourhood might be: 1) Do you spend a lot of time visiting parents and friends who live nearby? 2) Do you think prices are generally higher in your area than in other regions of the country? 3) Would you accept an interesting job elsewhere which would force you to move? The problem is that, most often, all people do not reply in accordance with the scale. The ordering is not respected and these individuals are unscalable.

With Q questions, a Guttman scale has $Q+1$ categories, ranging from none to all favourable responses. To these, we shall add an extra category, grouping together those who are not scalable. We then make the hypothesis that all of the responses to the Q questions for these unscalable individuals are independent. However, unscalable individuals may fall on the scale by chance, so that, in each of the first $Q+1$ categories, we have a mixture of two distinct populations answering according to the scale. We cannot disaggregate them with the available data.

The sum of the probabilities for the $Q+2$ categories must be one; represent them by π_k ($k = 0, \ldots, Q+1$). For the unscalable individuals in the last category, each question, i, has an independent probability, say ν_1^i, of reply 'yes' ($\nu_2^i = 1 - \nu_1^i$). Then, with $Q = 3$, we would have the mixture model

Table 5.6. Guttman scale for role conflict. (Stouffer and Toby, 1951)

Question				
1	2	3	4	Frequency
1	1	1	1	42
1	1	1	2	23
1	1	2	1	6
1	1	2	2	25
1	2	1	1	6
1	2	1	2	24
1	2	2	1	7
1	2	2	2	38
2	1	1	1	1
2	1	1	2	4
2	1	2	1	1
2	1	2	2	6
2	2	1	1	2
2	2	1	2	9
2	2	2	1	2
2	2	2	2	20

$$\begin{aligned}
\pi_{111} &= \pi_1 + \pi_0\nu_1^1\nu_1^2\nu_1^3 \\
\pi_{112} &= \pi_2 + \pi_0\nu_1^1\nu_1^2\nu_2^3 \\
\pi_{122} &= \pi_3 + \pi_0\nu_1^1\nu_2^2\nu_2^3 \\
\pi_{222} &= \pi_4 + \pi_0\nu_2^1\nu_2^2\nu_2^3
\end{aligned} \tag{5.6}$$

for individuals replying according to the scale, where the first term corresponds to scalable individuals and the second to the unscalable ones, and

$$\pi_{jkl} = \pi_0\nu_j^1\nu_k^2\nu_l^3 \tag{5.7}$$

for the remaining unscalable ones, those with responses not on the scale. We, then, eliminate those categories which are heterogeneous, i.e. contain both scalable and unscalable individuals — those on the scale by chance — described by Equation (5.6), by giving them zero weights, and fit a quasi-independence model for Equation (5.7).

If the parameter estimates have been standardized to sum to zero, the probabilities of Equation (5.7) are given by

$$\nu_j^i = \frac{1}{1 + \exp(2 * \mathsf{PE})}$$

where PE is the parameter estimate supplied by the software. The probability, π_0, of being unscalable can now be calculated directly from Equation (5.7) using any fitted value supplied by the software.

$$\pi_0 = \frac{n_{jkl}/n_{\bullet\bullet\bullet}}{\nu_j^1 \nu_k^2 \nu_l^3}$$

Because the estimates of π_{jkl} in Equation (5.6) are simply the observed relative frequencies for these categories, the $Q+1$ probabilities of replying according to the scale may be calculated by subtracting fitted from observed values and dividing by the total.

The model will be applied to a scale of four questions describing situations of conflict between personal obligations to a friend and societal obligations, given in Table 5.6. We note the large numbers replying according to the scale and the scattering of individuals off the scale, but also the combination (1212) which has a large frequency of response.

The analysis, using the model

$$\texttt{Q1} + \texttt{Q2} + \texttt{Q3} + \texttt{Q4}$$

with five frequencies weighted out, has a deviance 0.99 (10.99) with 6 d.f. The model fits very well; indeed, it is overfitted, as can be seen from the normal probability plot in Figure 5.6. The plot of Cook's distances, in Figure 5.7, indicates that the combinations, (1121) and (2212), as well as (1212), fit most poorly.

Thus, there are a large number of persons replying off the scale, with 24 in the one specific way (1212). This has already placed the original construction of the questions under suspicion. Removing either question 2 or question 3 does not really improve matters. Indeed, the results from our model indicate the probability of an unscalable reply to be 0.683. Probabilities of replying 'yes' to each question by chance are estimated as (0.766, 0.384, 0.442, 0.195), while the probabilities of replying according to the scale (0.177, 0.035, 0.026, 0.031, 0.048) do not change regularly, and are all small, again placing the scale in question.

With data following a well-constructed Guttman scale, the probability of replying by chance should be small so that the sum of probabilities of replying on the scale would be almost one. These probabilities would then indicate to which end of the scale individuals tend to lean.

5.6 Exercises

(1) Fingerprints on the right hand were classified as to the numbers of fingers with whorls and small loops (Plackett 1974, p. 22, from Waite):

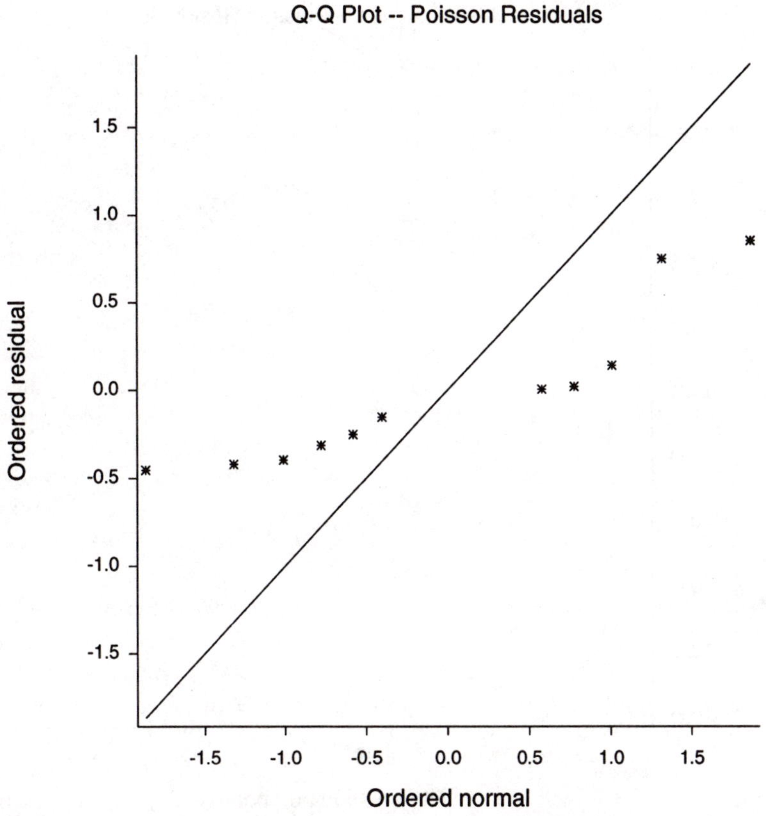

Fig. 5.6. Normal probability plot from the Guttman model of Equation (5.7) for the role conflict data of Table 5.6.

	Small loops					
Whorls	0	1	2	3	4	5
0	78	144	204	211	179	45
1	106	153	126	80	32	—
2	130	92	55	15	—	—
3	125	38	7	—	—	—
4	104	26	—	—	—	—
5	50	—	—	—	—	—

Develop a suitable model for these data.

(2) In a study of the relative frequencies of different kinds of sperm produced by male of the fly, *Drosophilus melanogaster*, carrying the translocation between the X and the fourth chromosome, known as $T(1,4)B^S$, the following results were obtained (Haberman, 1974a, p.

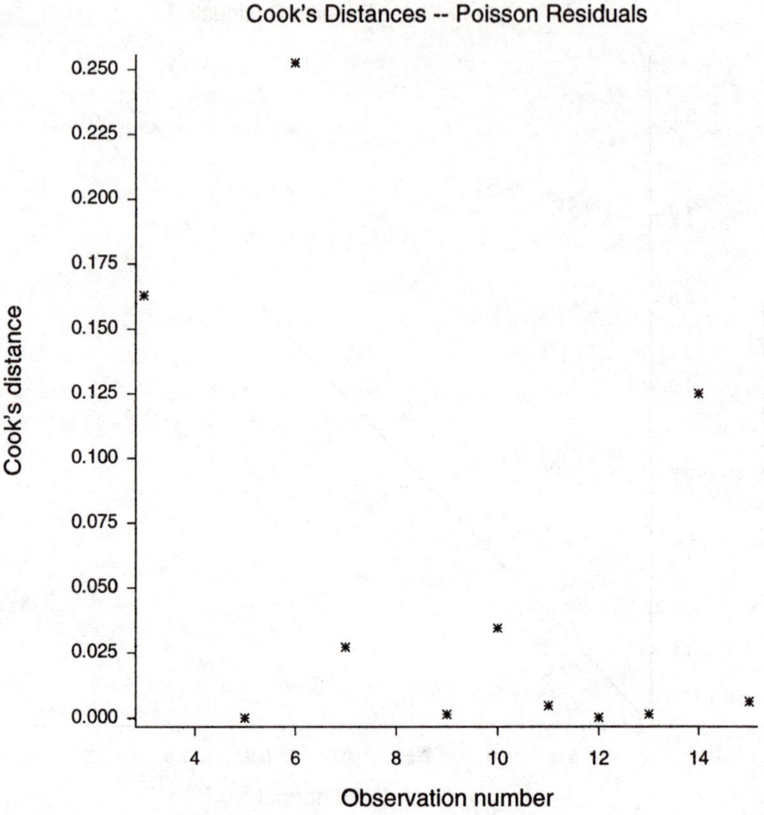

Fig. 5.7. Cook's distances from the Guttman model of Equation (5.7) for the role conflict data of Table 5.6.

229, from Novitski and Sandler):

Fourth chromosome	Type of Female	Male gamete type AB	AB'	$A'B$	$A'B'$
Marked	Y	847	—	1349	591
Unmarked		566	—	891	438
Marked	B	—	196	655	273
Unmarked		—	150	632	275

In the first homologue, A represents the translocation segment carrying the distal end of the X chromosome. If not present, then the fourth chromosome, denoted by A', must be present. Similarly, in the second homologue, B represents the segment of the translocation carrying the base of the X chromosome and B' the Y chromosome. Males carrying the translocation were crossed with attached X females carrying either a Y chromosome or a section B of the translocation. As seen in the table, two combinations were lethal. Develop an appropriate model for these data.

(3) At the beginning of the school year, boys in several secondary schools were asked how they would like to be remembered, as an athlete or as a brilliant student, as well as whether they would be playing football that autumn (Coleman, 1964, p. 404). The results for two schools are:

	Year							
	Freshman		Sophomore		Junior		Senior	
	Will play football							
Remembered as	Yes	No	Yes	No	Yes	No	Yes	No
	School 8							
Athlete	—	—	40	130	32	79	22	53
Student	—	—	12	117	8	78	6	62
	School 9							
Athlete	28	48	36	87	21	65	20	52
Student	3	54	3	49	2	35	6	38

Note that school 8 has no freshman year.

(a) Does the way boys wish to be remembered differ between schools?
(b) Does it vary over the years of schooling?
(c) How does it depend on athletic activity?

(4) Colour forms of the meadow spittlebug *Philaenus spumarius* were studied at two locations in Wisconsin, U.S.A. Many forms occur only for females. (Bishop *et al.*, 1975, p. 224, from Thompson)

	Male		Female	
	Location			
	A	B	A	B
populi	51	116	58	133
typica	123	303	91	289
trilineata	18	44	15	44
marginella	—	—	7	31
lateralis	—	—	5	13
flavicollis	—	—	1	2
albomaculata	1	1	2	7

Do the colour forms vary with sex or location?

(5) Christensen (1990, p. 340) gives the results of arthroscopic knee surgery:

Age	Sex	Type of injury	Result Excellent	Good	Fair–poor
11–30	M	Twist	21	11	4
31–50			32	20	5
51–91			20	12	5
11–30	F		3	1	0
31–50			6	5	2
51–91			6	3	1
11–30	M	Direct	3	2	2
31–50			2	4	4
51–91			0	0	0
11–30	F		0	1	1
31–50			0	0	0
51–91			1	2	3
11–30	M	Both	7	1	1
31–50			11	6	2
51–91			0	4	6
11–30	F		1	0	0
31–50			1	1	1
51–91			2	4	1
11–30	M	None	0	0	0
31–50			1	2	1
51–91			3	3	0
11–30	F		1	0	0
31–50			1	2	0
51–91			1	6	8

Find an appropriate model to describe how the result of surgery depends on age, sex, and the type of injury, taking care to account for the zero frequencies in the table and the ordinal variable describing the result.

(6) The following table gives the number of albinos in families of different sizes (Kocherlakota, and Kocherlakota, 1990, from Rao).

Number of albinos	Size of family 4	5	6	7
1	22	25	18	16
2	21	23	13	10
3	7	10	18	14
4	0	1	3	5
5	—	1	0	1
6	—	—	1	0
7	—	—	—	0

Does the proportion of albinos in a family depend on the family size?

(7) Initial and final ratings of hospitalized stroke patients were recorded

on entering and leaving hospital (Bishop and Fienberg, 1969). Patients who deteriorate are never discharged, so that certain cells are missing.

	Final				
Initial	A	B	C	D	E
A	5	—	—	—	—
B	4	5	—	—	—
C	6	4	4	—	—
D	9	10	4	1	—
E	11	23	12	15	8

Develop an appropriate model taking into account the ordering of the variables.

(8) The psychology study in Table 5.4 was repeated with each subject being asked to return a message most like that received. The results (Bishop *et al.*, 1975, p. 297, from Foa) were

	Returned					
Received	Love	Status	Inform.	Money	Goods	Services
Love	—	73	11	0	2	25
Status	69	—	22	11	3	6
Inform.	19	18	—	12	27	15
Money	0	18	9	—	66	18
Goods	7	6	23	61	—	14
Services	46	20	8	18	19	—

Check to see if a quasi-symmetry model fits these data.

(9) Genital displays were observed among pairs of squirrel monkeys in a colony of six animals (Bishop *et al.*, 1975, p. 203, from Ploog):

Active	Passive participant					
participant	1	2	3	4	5	6
1	—	1	5	8	9	0
2	29	—	14	46	4	0
3	0	0	—	0	0	0
4	2	3	1	—	38	2
5	0	0	0	0	—	1
6	9	25	4	6	13	—

(a) Are the displays of the active and passive participants independent?

(b) Is there symmetry between the two types of participant?

(10) The education of married couples participating in the 1972 General Social Survey in the U.S.A. was distributed as follows (Haberman, 1979, p. 521):

	Wife			
Husband	0–11	12	13–15	16+
0–11	283	141	25	4
12	82	180	43	14
13–15	20	104	43	20
16+	4	52	41	69

What is the relationship between the education of the two spouses?

(11) Children in Massachusetts, U.S.A., possessing a specific congenital anomaly were recorded, based on I. obstetric records, II. other hospital records, III. schools, IV. Massachusetts Department of Health, and V. Massachusetts Department of Mental Health (Bishop *et al.* 1975, p. 253):

Record	Frequency	Record	Frequency
YYYYY	2	NYYYY	0
YNYYY	3	NNYYY	0
YYNYY	8	NYNYY	23
YNNYY	5	NNNYY	30
YYYNY	2	NYYNY	0
YNYNY	5	NNYNY	3
YYNNY	18	NYNNY	34
YNNNY	36	NNNNY	83
YYYYN	5	NYYYN	3
YNYYN	1	NNYYN	2
YYNYN	25	NYNYN	37
YNNYN	22	NNNYN	97
YYYNN	1	NYYNN	1
YNYNN	4	NNYNN	4
YYNNN	19	NYNNN	37
YNNNN	27	NNNNN	—

Estimate the total number of children with such congenital anomalies in Massachusetts.

(12) Table 5.6 gave responses to four situations of role conflict. There, the questions placed the respondent in a conflict situation. Equal numbers of persons were asked similar questions where a friend was in the situation and where a third party was (Stouffer and Toby, 1951). These are given in the following table, along with those from Table 5.6:

Question				Self	Friend	Other
1	2	3	4			
1	1	1	1	42	37	35
1	1	1	2	23	31	17
1	1	2	1	6	6	9
1	1	2	2	25	15	26
1	2	1	1	6	5	3
1	2	1	2	24	29	27
1	2	2	1	7	6	3
1	2	2	2	38	25	32
2	1	1	1	1	2	3
2	1	1	2	4	4	5
2	1	2	1	1	3	2
2	1	2	2	6	4	5
2	2	1	1	2	3	0
2	2	1	2	9	23	20
2	2	2	1	2	3	3
2	2	2	2	20	20	26

(a) Check whether the same model can be fitted to all three types of question.

(b) Explain any anomalies.

6
Fitting distributions

Categorical data models have traditionally been used to study nominal, and ordinal, response variables. However, empirically, 'continuous' variables are also always observed as discrete categories, defined by the precision of the measuring instrument. Thus, log linear modelling techniques can be applied to such data, at least if the frequencies for the categories are reasonably large.

6.1 Exponential family

Let us now look at the data on recall of events in Table 1.1 in a different way. There, we took the occurrence or frequency of events as the response variable and time as the explanatory variable. Now we shall take $y_i = i$, the number of months to a recalled event for each individual, as our response variable, with an exponential distribution:

$$f(y_i; \lambda) = \lambda e^{-\lambda y_i}$$

However, our data are grouped into one-month intervals, so that we have

$$\begin{aligned}\Pr\left(y_i - \frac{\Delta_i}{2} < Y_i \le y_i + \frac{\Delta_i}{2}; \lambda\right) &= \int_{y_i - \frac{\Delta_i}{2}}^{y_i + \frac{\Delta_i}{2}} \lambda e^{-\lambda y_i} dy_i \\ &\doteq \lambda e^{-\lambda y_i} \Delta_i\end{aligned}$$

where Δ_i is one month. Taking logarithms, we obtain

$$\begin{aligned}\log(n_\bullet \pi_i) &= \log(n_\bullet \lambda \Delta_i) - \lambda y_i \\ &= \log(\Delta_i) + \beta_0 + \beta_1 y_i\end{aligned}$$

where $\pi_i = \Pr(y_i - \frac{\Delta_i}{2} < Y_i \le y_i + \frac{\Delta_i}{2})$, $\beta_0 = \log(n_\bullet \lambda)$, $\beta_1 = -\lambda$, and $\log(\Delta_i)$ is an offset, although it can be ignored because $\Delta_i = 1$. This is identical to our Poisson regression model in Section 1.2 above, with $\nu_i = n_\bullet \pi_i$. Thus, our analysis there can be interpreted as fitting an exponential distribution to the times in those data or, equivalently, a Poisson distribution to the frequencies.

The exponential distribution is one of the simplest members of the exponential family:

$$f(y_i;\boldsymbol{\theta}) = \exp\left[\boldsymbol{\theta}^T\mathbf{t}(y_i) + c(\boldsymbol{\theta}) + d(y_i)\right]$$

where $\mathbf{t}(y_i)$ is the vector of sufficient statistics for the canonical parameters, $\boldsymbol{\theta}$. For any empirically observed data, the precision of measurement is finite, so that the probability function, $\pi_i = \Pr(y_i - \frac{\Delta_i}{2} < Y_i \leq y_i + \frac{\Delta_i}{2})$, must be used, even if the variable is theoretically continuous. We thus apply the same approximation to the integral, as above, for continuous variables. If we again take logarithms, we may note that

$$\log(n_{\bullet}\pi_i) = \boldsymbol{\theta}^T\mathbf{t}(y_i) + c(\boldsymbol{\theta}) + d(y_i) + \log(n_{\bullet}\Delta_i)$$

which is linear in $\mathbf{t}(y_i)$. Because $d(y_i) + \log(n_{\bullet}\Delta_i)$ contains no unknown parameters, it can be used as an offset, so that we have

$$\log(n_{\bullet}\pi_i) = \beta_0 + \boldsymbol{\beta}^T\mathbf{t}(y_i)$$

where $\boldsymbol{\beta} = \boldsymbol{\theta}$ and $\beta_0 = c(\boldsymbol{\theta})$.

Thus, for frequency data, any distribution from the exponential family can be represented as a Poisson log linear regression model (Lindsey and Mersch, 1992). Other distributions can also be so represented, but the regression equation will no longer be linear.

In the Poisson regression, the 'explanatory' variables are the sufficient statistics of the appropriate member of the exponential family. The model terms for common distributions include

Distribution	Sufficient statistics	Offset
Uniform		
Geometric	y_i	
Poisson	y_i	$-\log(y_i!)$
Binomial	y_i	$\log\binom{n_i}{y_i}$
Normal	y_i, y_i^2	
Log normal	$\log(y_i), \log^2(y_i)$	
Inverse Gauss	y_i, y_i^{-1}	$-1.5\log(y_i)$
Exponential	y_i	
Pareto	$\log(y_i)$	
Gamma	$y_i, \log(y_i)$	

Thus, for example, for the Poisson distribution with mean, λ, we fit

$$\log(\nu_i) = \beta_0 + \beta_1 y_i$$

with offset $\log(n_{\bullet}/y_i!)$, so that $\beta_0 = -\lambda$ and $\beta_1 = \log(\lambda)$. For the normal distribution, with mean, λ, and variance, σ^2, we fit

$$\log(\nu_i) = \beta_0 + \beta_1 y_i + \beta_2 y_i^2$$

so that $\beta_1 = \lambda/\sigma^2$ and $\beta_2 = -1/(2\sigma^2)$, the canonical parameters, and the offset is $\log(\Delta_i)$.

To recapitulate, we have the multinomial likelihood function

$$\mathrm{L}(\boldsymbol{\pi}) \propto \prod \pi_i^{n_i}$$

where n_i is the frequency in category i and

$$\begin{aligned}
\pi_i &= F(y_i; \boldsymbol{\theta}) \qquad y \text{ discrete} \\
&= \int_{x_i - \frac{\Delta_i}{2}}^{x_i + \frac{\Delta_i}{2}} f(y_i; \boldsymbol{\theta}) dy \qquad y \text{ continuous} \\
&\doteq f(y_i; \boldsymbol{\theta}) \Delta_i
\end{aligned}$$

As we saw in Section 1.1, for fixed $n_\bullet$, this is equivalent to a Poisson likelihood

$$\prod \frac{\nu_i^{n_i} \mathrm{e}^{\nu_i}}{n_i!}$$

where $\log(\nu_i) = \boldsymbol{\theta}^T \mathbf{t}(y_i) + c(\boldsymbol{\theta}) + d(y_i) + \log(n_\bullet \Delta_i)$ for the exponential family.

As we have seen, in log linear models for categorical data, conditioning on the total number of observations in the Poisson likelihood ensures that the total multinomial probability of all categories included in the model equals unity. This is accomplished by retaining the intercept in the model to fix the marginal total; we now see that this intercept is $\beta_0 = c(\boldsymbol{\theta})$. The latter is just the normalizing constant of the exponential family. Thus, this fitting procedure ensures that we have a proper probability distribution which sums to one over the observed categories. It also means that we do not even have to know the analytic form of $c(\boldsymbol{\theta})$ in order to fit the model.

In Chapter 5, we distinguished between sampling and structural zeros. Sampling zeros occur for categories which did not happen to be observed in the given data. Structural zeros are categories which are impossible for the given data. We saw that these must be treated differently: categories with sampling zeros are included in a log linear model, while those for structural zeros are not.

Consider a response variable with some range such as $y > 0$. This means that an infinite number of categories, $(y_i - \frac{\Delta_i}{2}, y_i + \frac{\Delta_i}{2}]$, is possible, but only a few have been observed. The others are sampling zeros and should be included in the model. However, most have extremely small probabilities and their exclusion does not greatly affect the parameter estimates. Thus,

Table 6.1. Employment durations of Post Office staff. (Burridge, 1981)

	Grade			Grade	
Months	1	2	Months	1	2
1	22	30	13	0	1
2	18	28	14	0	0
3	19	31	15	0	0
4	13	14	16	1	1
5	5	10	17	1	1
6	6	6	18	1	0
7	3	5	19	3	2
8	2	2	20	1	0
9	2	3	21	1	3
10	1	0	22	0	1
11	0	0	23	0	1
12	1	1	24	0	0

sampling zeros can be added for enough categories to obtain any required precision.

If the people interviewed for Table 1.1 had not been limited to events in the previous 18 months, we might have had responses for months further back in the past. To fit the exponential distribution, we would then first need to extend the vector with zeros past the last month in which there was a response, perhaps to 25 or 30 months in all, before fitting the model as we did in Section 1.2.

6.2 Comparing models

Once we begin to treat probability distributions in the log linear model context, there is no reason why we should restrict ourselves to 'explanatory' variables which are the sufficient statistics for only one distribution. We saw that such statistics include y_i, y_i^2, $\log(y_i)$, $\log^2(y_i)$, y_i^{-1}. Any number of these functions, and others, can be included in a log linear model for such frequency data and standard regression techniques can be used to determine which can be eliminated. Thus, for example, to compare log normal and gamma distributions, the statistics, y_i, $\log(y_i)$, $\log^2(y_i)$ would be used. If y_i can be eliminated, we have a log normal distribution, whereas, if $\log^2(y_i)$ is unnecessary, we have a gamma distribution. This may also mean that some more complex combination proves necessary, not corresponding to any distribution commonly used.

Burridge (1981) fits a gamma distribution to the employment durations of two grades of staff recruited to the Post Office in the first quarter of 1973, as shown in Table 6.1. The log linear model for this distribution, with y_i and $\log(y_i)$,

$$\texttt{GRADE} + \texttt{Y} + \texttt{LOGY} \tag{6.1}$$

gives a deviance of 76.98 (84.98) with 44 d.f. (Sampling zeros have not been added because it is not clear that observations were made for more than 24 months.) If we also try a number of other statistics, we find that a model with y_i, $\log(y_i)$, y_i^{-1}, y_i^{-2}

$$\texttt{GRADE} + \texttt{Y} + \texttt{LOGY} + \texttt{RECY} + \texttt{REC2Y} \tag{6.2}$$

has a deviance of 39.79 (51.79) with 42 d.f. Such a distribution will have an analytically complex normalizing constant. As seen in Figure 6.1, this yields a survivor function which, at first, drops more quickly than the gamma, but then stays at a higher level for long durations. We also see that the observed values stay fairly flat in the middle period, until about 15 months, and that the more complex model is trying to take this into account.

If categorical explanatory variables are available, we have generalized linear models and the data will take the form of a multi-dimensional contingency table. Again, standard log linear techniques can be applied. The usual marginal totals are fixed by including the correct combination of explanatory variables in the minimal model, as they were in Equation (2.9), to ensure that the response distributions are suitably normalized. The same sufficient statistics are used, but now in interaction with the explanatory variables. If the interaction terms are not necessary in the model, the corresponding explanatory variable can be eliminated: the response does not depend on it.

In our example, the change in deviance for adding a difference in grade

$$\texttt{GRADE} * (\texttt{Y} + \texttt{LOGY} + \texttt{RECY} + \texttt{REC2Y})$$

to the model is 0.80 (AIC 58.99) with 4 d.f. Note that here we are fitting a distribution with four completely different parameter values for each grade.

We now have a further generalization: the coefficient for a sufficient statistic may be zero in some categories of an explanatory variable and not in others. This would mean that we have different distributions for different categories of the explanatory variables. Thus, in one model, we can fit several distributional forms. This might arise, for example, in a mortality study, where test and control sub-populations have *functionally* different hazard functions, and not simply different parameter values.

6.3 Intractable normalizing constants

A number of the more complex models in the previous section, with several sufficient statistics fitted, including that finally chosen for the data of Table 6.1, have distributions whose normalizing constant does not have a simple

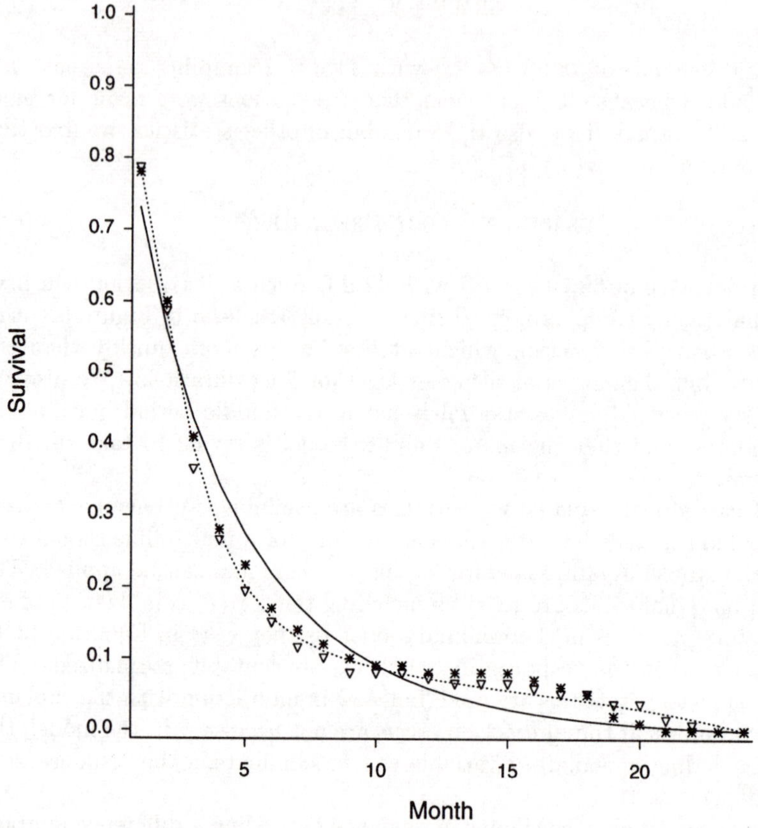

Fig. 6.1. Fitted gamma (6.1 — solid) and more complex (6.2 — dotted) survival curves, with the observed values, for the employment data in Table 6.1.

form. These create difficulties in estimating parameters or plotting the likelihood function in the usual way, which has meant that they have been little used.

One class of distributions with such intractable normalizing constants is the double exponential family (Efron, 1986; Aitkin, 1994):

$$f(y;\theta,\psi) = c(\theta,\psi)\sqrt{\psi}[f(y;\theta)]^{\psi}[f(y;y)]^{1-\psi}$$

where $f(y;\theta)$ is some member of the one parameter exponential family. It can easily be seen that the resulting distribution is a member of the two parameter exponential family. Such models are especially useful when the one parameter model cannot account for all of the variability in the data, known as overdispersion; see Chapter 10.

Table 6.2. Frequency of males in 6115 families with 12 children in Saxony. (Sokal and Rohlf, 1969, p. 80, from Geissler)

Males	Families
0	3
1	24
2	104
3	286
4	670
5	1033
6	1343
7	1112
8	829
9	478
10	181
11	45
12	7

Thus, for example, we obtain the double Poisson distribution

$$f(y;\nu,\psi) = c(\nu,\psi)\frac{\sqrt{\psi}}{\mathrm{e}^{\psi\nu}y!}\left(\frac{y}{\mathrm{e}}\right)^{y}\left(\frac{\nu\mathrm{e}}{y}\right)^{y\psi}$$

where the sufficient statistics, y and $y[1-\log(y)]$, would be fitted with offset, $-\log(y!)$, and the double binomial distribution

$$f(y;\pi,\psi) = c(\pi,\psi)\binom{n}{y}\frac{y^{y}(n-y)^{n-y}}{n^{n}}\frac{n^{n\psi}}{y^{y\psi}(n-y)^{(n-y)\psi}}\pi^{y\psi}(1-\pi)^{(n-y)\psi}$$

with sufficient statistics, y and $-y\log[y/(n-y)]$, and offset, $\log\binom{n}{y}$. In neither case can $c(\cdot)$ be simply written down.

Geissler recorded a classical set of data on sex ratios, from hospital records in Saxony in the nineteenth century. Those for families of 12 children are shown in Table 6.2. The probability of a male child, estimated from these data, is 0.52. With the binomial distribution, fitting `Y` with offset, $\log\binom{n}{y}$, the deviance is 97.01 (101.01) with 11 d.f., a poor fit, while, for the double binomial,

$$\texttt{Y} + \texttt{YLOGITY} \tag{6.3}$$

it is 30.28 (36.28) with one less degree of freedom. Although the latter is not a good fit, it is much better than the binomial. The fitted values for the two models, with the residuals, are given in Table 6.3. The binomial model shows a particular pattern in the residuals: negative values in the

Table 6.3. Fitted values and residuals from the binomial and double binomial (6.3) models for the sex of chlidren data of Table 6.2.

		Binomial		Double binomial	
Males	Observed	Fitted	Residual	Fitted	Residual
0	3	0.9	2.14	1.9	0.81
1	24	12.1	3.42	18.9	1.16
2	104	71.8	3.79	94.8	0.95
3	286	258.5	1.71	299.0	−0.75
4	670	628.1	1.68	657.6	0.48
5	1033	1085.2	−1.59	1059.9	−0.83
6	1343	1367.3	−0.66	1285.4	1.61
7	1112	1265.6	−4.32	1187.0	−2.18
8	829	854.2	−0.86	835.6	−0.23
9	478	410.0	3.36	444.7	1.58
10	181	132.8	4.18	176.4	0.35
11	45	26.1	3.71	52.7	−1.06
12	7	2.3	3.04	1.0	5.90

centre and positive ones in both tails. This is an example of overdispersion, which we shall study in Chapter 10. The binomial and double binomial models are plotted in Figure 6.2, again confirming the superiority of the latter. It fits the tails more closely, at the expense of a lower estimate for families with equal numbers of the two sexes.

Complicated normalizing constants in distributions are a fairly common occurrence. We shall now look at some other situations in which they can arise.

6.4 Multivariate distributions

If more than one response variable is present, multivariate distributions may be fitted. These will often have intractable normalizing constants, as in the previous section. For log linear models, these will have exponential family conditional distributions (Arnold and Strauss, 1991).

Let us look at a study of the distribution of income and wealth in Denmark in 1974, as given in Table 6.4. Here, there are two responses but no explanatory variables, so that the analysis is relatively simple. However, the wealth and income are fairly crudely grouped so that any results can only be approximate. We take values (1, 25, 100, 225, 450) for wealth (the first value arbitrarily set to unity, instead of zero so that logarithms can be used; the last is also arbitrary), with $\boldsymbol{\Delta}^T = (1, 50, 100, 150, 300)$, and (20, 50, 70, 95, 200) for income, with $\boldsymbol{\Delta}^T = (40, 20, 20, 30, 180)$. Obviously, the values chosen for the last category of each variable are rather arbitrary.

We shall first only consider a few standard bivariate distributions: the

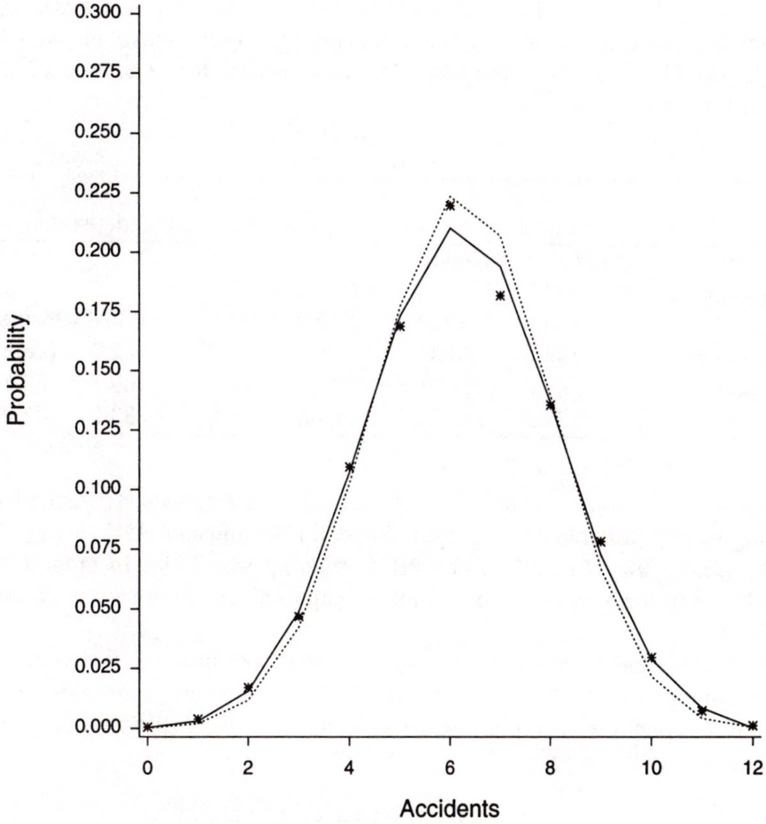

Fig. 6.2. Fitted binomial (dotted) and double binomial (6.3 — solid) distributions, with the observed values, for the sex of children data in Table 6.2.

Table 6.4. Distribution of income and wealth (1000 DKr.) in Denmark, 1974. (Andersen, 1991, p. 350)

	Wealth				
Income	0	1–50	50–150	150–300	300+
0–40	292	126	22	5	4
40–60	216	120	21	7	3
60–80	172	133	40	7	7
80–110	177	120	54	7	4
110+	91	87	52	24	25

Table 6.5. Deviances (AIC) for bivariate distributions for the income and wealth of Table 6.4. (The normal, log normal, inverse Gauss, and gamma models have 20 d.f. for the independent, 19 for dependent, and 18 for the doubly dependent models. The exponential has 2 more and the uniform 4 more.)

	Independent	Dependent	Doubly dependent
Uniform	13761.0 (13763.0)		
Exponential	5347.7 (5353.7)	4781.2 (4789.2)	
Normal	3130.6 (3140.6)	3019.0 (3031.0)	2635.8 (2649.8)
Log normal	193.3 (203.3)	76.2 (88.2)	52.3 (66.3)
Inverse Gauss	201.8 (211.7)	96.4 (108.4)	69.1 (83.1)
Gamma	179.6 (189.6)	64.5 (76.5)	33.2 (47.2)

exponential, gamma, normal, log normal, and inverse Gauss, instead of looking at any possible combination. We call the sufficient statistics `W`, `I`, `LOGW`, `LOGI`, `W2`, `I2`, `RECW`, and `RECI`, where `I` stands for income, `W` for wealth, a two for a square and `REC` for a reciprocal. No factor variables are involved.

The null model, which represents the bivariate uniform distribution, has a deviance of 13 761 with 24 d.f. Other results are summarized in Table 6.5. Thus, for example, the usual bivariate normal distribution, with dependence, is fitted by

$$\mathtt{W} + \mathtt{W2} + \mathtt{I} + \mathtt{I2} + \mathtt{W} \cdot \mathtt{I}$$

while what is called doubly dependent in the table has the model

$$\mathtt{W} + \mathtt{W2} + \mathtt{I} + \mathtt{I2} + \mathtt{W} \cdot \mathtt{I} + \mathtt{W2} \cdot \mathtt{I2}$$

In the same way, a doubly dependent bivariate gamma distribution is defined by

$$\mathtt{W} + \mathtt{LOGW} + \mathtt{I} + \mathtt{LOGI} + \mathtt{W} \cdot \mathtt{I} + \mathtt{LOGW} \cdot \mathtt{LOGI}$$

Of the models listed, the doubly dependent gamma, with 18 d.f., is best. As might be expected, both of the dependence parameters are positive: 0.000 030 9 for `W·I` and 0.113 for `LOGW·LOGI`. If we look at more complex distributions, such as those we used in Section 6.2, we find the model

$$\begin{aligned}\mathtt{W} + \mathtt{LOGW} + \mathtt{LOGW2} + \mathtt{I} + \mathtt{LOGI} + \mathtt{LOGI2} + \mathtt{W} \cdot \mathtt{I} + \mathtt{LOGW2} \cdot \mathtt{LOGI2}\\ + \mathtt{W} \cdot \mathtt{LOGW} + \mathtt{W} \cdot \mathtt{LOGI} + \mathtt{I} \cdot \mathtt{LOGW}\end{aligned}$$

Table 6.6. Numbers of occupants in houses replying to a postal survey. (Lindsey and Mersch, 1992)

Occupants	Houses
1	436
2	133
3	19
4	2
5	1
6	0
7	1

with a deviance of 14.14 (38.14) and 13 d.f. In fact, `LOGW2` and `LOGI2` can be eliminated, with little change in deviance, but the model would no longer be hierarchical. However, it is questionable if the added complexity, and lack of interpretability, of either of these models, is worth the reduction in deviance, as compared to the bivariate distribution with gamma conditionals.

It is interesting to note that the usual independence model for two-way tables has a deviance of 167.99 (185.99) with 16 d.f., similar to that for the independent bivariate gamma, with 179.60 (189.60) and 20 d.f., and much worse than the dependent gamma, with 33.22 (47.22) and 18 d.f. Introduction of any dependence structure into that standard log linear model would involve further loss of degrees of freedom.

In this example, we have retained a model where both conditional distributions have the same form, although with different parameter values. In general, this need not be the case. For example, the two variables defining a table might be length of marriage and number of children. Then a possible bivariate distribution could have gamma and Poisson conditional distributions. If there are explanatory variables, different distributions may also be found under different (combinations of) conditions, as suggested in Section 6.2.

6.5 Truncated distributions

So far, we have only considered sampling zeros in our probability distribution. Suppose, now, that we treat some zero categories (with reasonably large probabilities) as structural zeros and leave them out. In this way, we are fitting a truncated distribution. With our log linear approach, this can be done even more easily than for complete distributions. All of the techniques described above can still be used. (In fact, we fitted a truncated exponential distribution above in Sections 1.2 and 6.1.)

As an example, let us look at data obtained from a postal survey, given in Table 6.6. Obviously, houses with zero occupants could not reply to the postal questionnaire, so that a truncated distribution, without the zero

Table 6.7. Fitted values and residuals from the truncated Poisson distribution (6.4) for the postal survey data of Table 6.6.

Occupants	Observed	Fitted	Residuals
1	436	437.6	−0.078
2	133	126.2	0.607
3	19	24.3	−1.067
4	2	3.5	−0.800
5	1	0.4	0.940
6	0	0.0	−0.197
7	1	0.0	17.642

category, is appropriate. For these counts of numbers of people in houses, the truncated Poisson distribution

$$\Pr(y_i;\lambda) = \frac{e^{-\lambda}\lambda^{y_i}}{(1-e^{-\lambda})y_i!} \tag{6.4}$$

might be a suitable hypothesis.

We extend the vector of frequencies to length ten by adding zeros. Then, the number of occupants is fitted as an 'explanatory' variable with the usual Poisson offset. The resulting deviance is 12.56 (16.56) with 8 d.f., although there are four sampling zeros in the table. The parameter estimates are $\hat{\beta}_0 = 6.632$ and $\hat{\beta}_1 = -0.5505$. From the latter parameter, we have $\lambda = \exp(\beta_1)$, while, from the former, we have $\lambda = \log(e^{-\beta_0}+1)$. Both calculations yield $\hat{\lambda} = 0.577$ for this truncated Poisson distribution. The distribution is plotted in Figure 6.3, revealing a good fit. Inspection of the residuals in Table 6.7 shows, as might be expected, slight underestimation of households of size two. There are more couples than might be expected from the randomness of the Poisson distribution.

6.6 Mixture distributions

A finite mixture distribution is the weighted sum of several simple distributions:

$$\Pr(y) = \sum \pi_i \Pr_i(y)$$

with $\sum \pi_i = 1$. Often, only two distributions are involved. Such models could be fitted by our methods, using an identity link, but the regression equation would be non-linear.

One special case is of particular importance: we have a simple distribution, except that one category contains a mixture of two populations. For example, events may follow a Poisson distribution, but certain, unidentifiable, individuals may not be able to have events. The zero category will

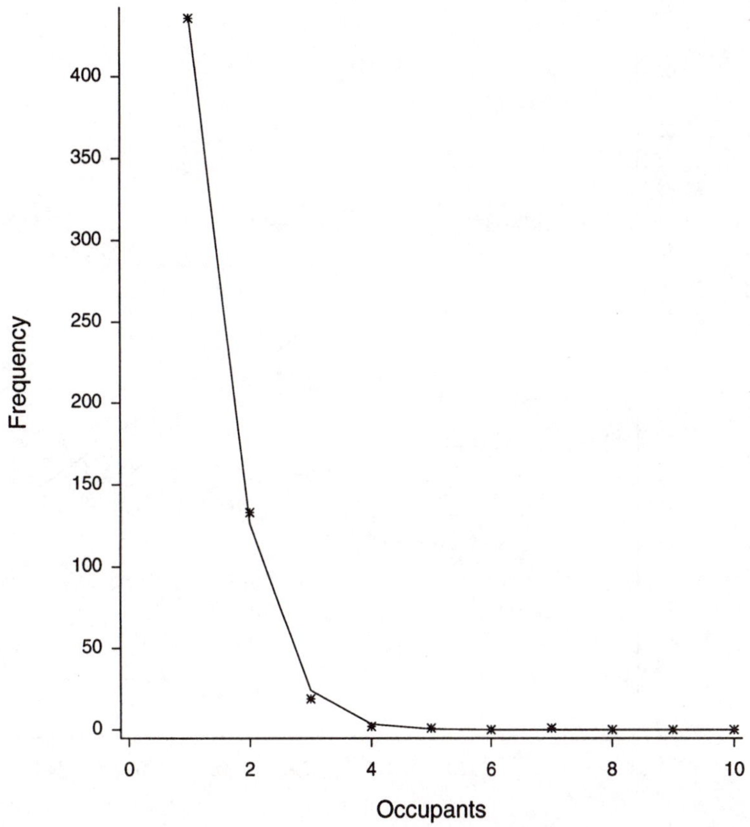

Fig. 6.3. Observed frequencies and fitted truncated Poisson distribution (6.4) for the postal survey data in Table 6.6.

be a mixture of cases not happening to have an event and those not able to have one. If this category is structurally eliminated, for example by weighting it out during the fit, the number of cases not happening to have an event can subsequently be estimated from the model. This is similar to what we did in Section 5.5 for Guttman scales. Note, however, that such a procedure does not guarantee that the estimated number without an event is less than the total observed number.

In the early detection of cancer, the study of genetic lesions, as indicated, for example, by micronuclei counts, is an important procedure. One of the most common factors causing such damage is some chemical agent, such as ethylene oxide. However, in such cases, the samples obtained may involve mixtures of affected and unaffected cells. Thus, if the counts of micronuclei in the affected cells are assumed to follow a Poisson distribution,

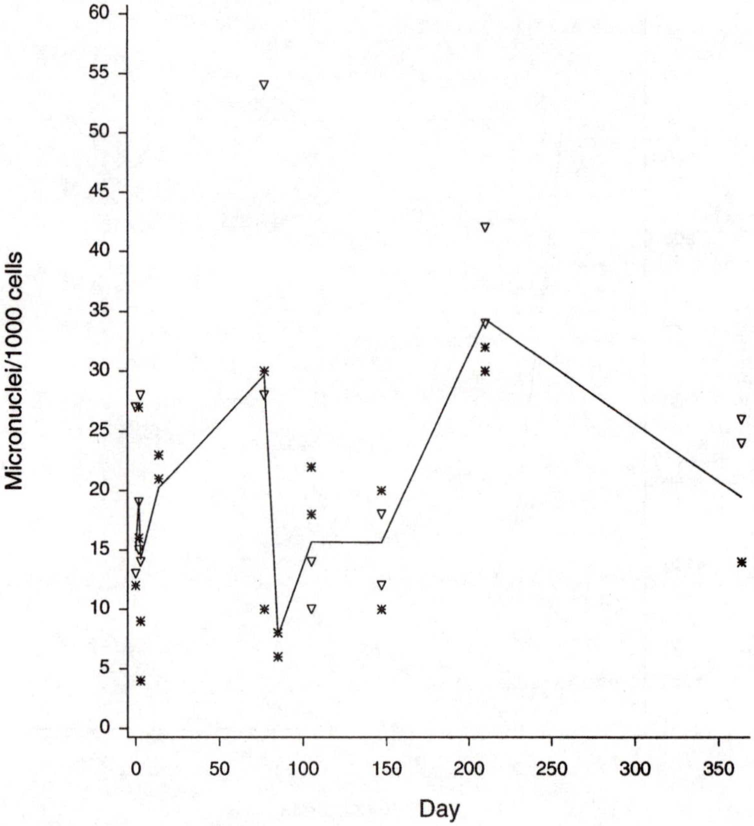

Fig. 6.4. Observed numbers of micronuclei each day for the data in Table 6.8, with the fitted double Poisson model of Equation (6.6), quadratic in days (stars: Subject 1; triangles: Subject 2).

an estimate of the proportion of affected cells can be obtained from a log linear model representation of a truncated Poisson distribution.

Data, given in Table 6.8, were available for two slide cultures per day from two subjects who were followed for a year after exposure to ethylene oxide. The two cases worked together at the University of Liège, Belgium, hospital, and continued to do so during the year of followup. However, as can be seen in the table, about 99% of cells examined contain no micronuclei. We have a contingency table for each subject, of sizes 20×6 and 15×6. The numbers of micronuclei are plotted for each subject in Figure 6.4.

We first fit the Poisson mixture model for independence,

Table 6.8. Micronuclei counts for two subjects over one year after exposure to ethylene oxide. (C. Laurent)

Day	Number of cells	Total number of micronuclei	Micronuclei counts 0	1	2	3	4	5
Subject 1								
0	1000	12	990	9	0	1	0	0
0	1000	12	992	7	0	0	0	1
2	1000	16	984	16	0	0	0	0
2	1000	27	977	20	2	1	0	0
3	1000	9	991	9	0	0	0	0
3	1000	4	996	4	0	0	0	0
14	1000	23	982	14	3	1	0	0
14	1000	21	983	14	2	1	0	0
77	1000	10	990	10	0	0	0	0
77	500	15	488	9	3	0	0	0
85	500	3	497	3	0	0	0	0
85	500	4	496	4	0	0	0	0
105	500	11	490	9	1	0	0	0
105	500	9	492	7	1	0	0	0
147	500	5	495	5	0	0	0	0
147	500	10	490	10	0	0	0	0
210	500	15	485	15	0	0	0	0
210	500	16	485	14	1	0	0	0
364	500	7	493	7	0	0	0	0
364	500	7	493	7	0	0	0	0
Subject 2								
0	1000	13	988	11	1	0	0	0
0	1000	27	974	25	1	0	0	0
2	1000	19	983	15	2	0	0	0
2	1000	15	985	15	0	0	0	0
3	1000	14	987	12	1	0	0	0
3	1000	28	974	24	2	0	0	0
77	500	27	479	18	1	1	1	0
77	500	14	487	12	1	0	0	0
105	500	7	493	7	0	0	0	0
105	500	5	496	3	1	0	0	0
147	500	9	491	9	0	0	0	0
147	500	6	495	4	1	0	0	0
210	500	21	481	17	2	0	0	0
210	500	17	485	13	2	0	0	0
364	500	13	488	11	1	0	0	0
364	500	12	489	10	1	0	0	0

$$\mathtt{Y} + \mathtt{SUBJECT} * \mathtt{DAY} * \mathtt{CULTURE}$$

with zero counts (not zero frequencies!) weighted out and the log of the number of cells observed included in the offset, where Y is the micronuclei count. We obtain a deviance of 77.18 (151.18) with 143 d.f., although this has no absolute meaning for goodness of fit because of the large number of zero frequencies. Including a factor variable for differences among days,

$$\mathtt{Y} * \mathtt{DAY} + \mathtt{SUBJECT} * \mathtt{DAY} * \mathtt{CULTURE} \qquad (6.5)$$

lowers the deviance to 62.61 (154.61) with 134 d.f., providing minimal indication of variation among days. Adding the difference between cultures or subjects hardly changes the deviance. Linear and quadratic terms for days do not improve the model either; the latter has a deviance of 74.59 (152.59) with 141 d.f.

If we try the double Poisson distribution of Section 6.3,

$$\mathtt{Y} + \mathtt{YLOGY} + \mathtt{SUBJECT} * \mathtt{DAY} * \mathtt{CULTURE}$$

we obtain a deviance of 62.59 (138.59) with 142 d.f. for the model with independence, a considerably better fit than the corresponding Poisson distribution. It is rather surprising that overdispersion, indicated by the need to include the term, YLOGY, can be detected in these data, although we are not using the zero frequencies to fit the model. If we now fit a quadratic model for the days,

$$(\mathtt{Y} + \mathtt{YLOGY}) * (\mathtt{LDAY} + \mathtt{QDAY}) + \mathtt{SUBJECT} * \mathtt{DAY} * \mathtt{CULTURE} \qquad (6.6)$$

the deviance is 56.54 (140.54) with 138 d.f., a somewhat poorer model, but certainly better than the Poisson model with a factor variable for days, and hence perhaps a model worth looking at. Removal of any one of the four interaction terms between the sufficient statistics and days brings the deviance back up to about that of the double Poisson model without days. Thus, both the mean number of abnormalities and their dispersion may be changing over time. The fitted numbers of micronuclei/1000 cells are plotted in Figure 6.4. This is an example of a regression model, first suggested in Lindsey (1974b), where both the location and dispersion parameters are changing with the explanatory variables.

From this model, we can obtain an estimate of the proportion of unaffected cells as it evolves over time. However, several frequencies of zero counts are estimated to be slightly greater than those observed. Setting these proportions to one, we obtain the graph plotted in Figure 6.5. The clear pattern in this graph contrasts with the disarray of Figure 6.4 for the total numbers of micronuclei observed over time. The present plot seems

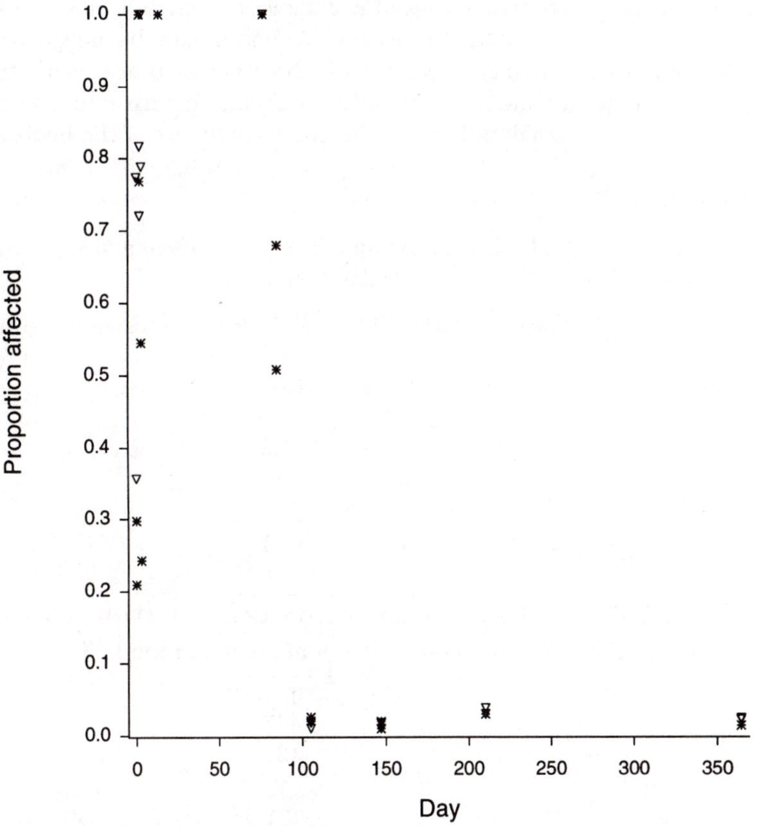

Fig. 6.5. Estimated proportion of affected cells from the model of Equation (6.6) for the micronuclei data in Table 6.8. (Stars: Subject 1; triangles: Subject 2)

to indicate that many, if not all, cells were contaminated up until about day 85, after which the proportion decreased to normal (as known from examining uncontaminated individuals). This appears to be the turnover time for new cells to be generated. Unfortunately, the times between observations are long in this middle period, so that no precise information is available on the turning point.

It is interesting to note that, if these proportions are estimated from the Poisson model of Equation (6.5) with days as a factor variable, they jump from around 10% up until day 105 to about 25% starting at day 147. This difference between the two models occurs because the Poisson model does not take into account the overdispersion, which is changing over time.

In the sex of children example of Section 6.3 and the postal survey example of Section 6.5, frequencies of counts were being modelled. In the same way, in this last example, the events may not, in fact, be independent, because they are occurring to the same subjects over time. With such irregular observation times, this would be difficult to take into account. Such examples of count data lead us into the second part of the book.

6.7 Exercises

(1) The following table shows the number of accidents in a year for 9461 Belgian drivers (Gelfand and Dalal, 1990):

Accidents	Drivers
0	7840
1	1317
2	239
3	42
4	14
5	4
6	4
7	1

Try to find a suitable probability distribution to fit these data.

(2) Skellam (1948) gives a distribution of accidents in the U.K.:

Accidents	Frequency
0	447
1	132
2	42
3	21
4	3
5	2
6	0

(a) Look for a suitable probability distribution to fit these data.
(b) Is the result similar to that for the previous question?

(3) The numbers of occupants in passenger cars were observed at an intersection in Los Angeles, U.S.A., during 40 minutes on the morning of 24 March, 1959 (Derman *et al.* 1973, p. 278):

Occupants	Cars
1	678
2	227
3	56
4	28
5	8
6+	14

Find a suitable distribution to fit these data. This will require some careful thought.

(4) In Table 6.2, we studied the distribution by sex in families of 12 children. The following table (Fisher, 1958, from Geissler) gives the results from the same study for families of eight children.

Males	Families
0	215
1	1485
2	5331
3	10649
4	14959
5	11929
6	6678
7	2092
8	342

(a) Is a binomial distribution suitable in this case?
(b) Is overdispersion present?

(5) Numbers of accidents to men working in a soap factory were recorded over a five-month period (Irwin, 1975):

Accidents	Men
0	239
1	98
2	57
3	33
4	9
5	2
6	2
7	1
8	0
9	4
10	1
11	0
12	0
13	1

Look for an appropriate distribution to fit these data.

(6) The table below gives the number of fire losses per year from 1950 to 1973 for a major university (Aiuppa, 1988).

Losses	Years
0	0
1	3
2	7
3	2
4	5
5	1
6	3
7	2

Look for an appropriate distribution to fit these data.

(7) Numbers of units purchased in four weeks were recorded for a particular brand of item (Derman *et al.*, 1973, p. 291):

Units	Consumers
0	1671
1	43
2	19
3	9
4	2
5	3
6	1
7	0
8	0
9	2
10	0

Look for a suitable probability distribution for these data.

(8) The table below shows the days between coal-mining disasters in Great Britain from 1851 to 1962 (Lindsey, 1992, pp. 66–67, from Jarrett, 1979).

Days	Disasters	Days	Disasters
0–20	28	140–200	17
20–40	20	200–260	16
40–60	17	260–320	11
60–80	11	320–380	13
80–100	14	380–440	3
100–120	6	440–500	4
120–140	13	> 500	17

Look for a suitable distribution to fit these data.

(9) The table below gives the monthly salaries (dollars) of female mathematics graduates practising mathematics or statistics (Zelterman, 1987).

Monthly salary	Number	Monthly salary	Number
1051–1150	1	2651–2750	5
1151–1250	1	2751–2850	6
1251–1350	6	2851–2950	4
1351–1450	3	2951–3050	4
1451–1550	4	3051–3150	5
1551–1650	3	3151–3250	1
1651–1750	9	3251–3350	1
1751–1850	6	3351–3450	4
1851–1950	5	3451–3550	1
1951–2050	16	3551–3650	2
2051–2150	4	3651–3750	0
2151–2250	11	3751–3850	2
2251–2350	6	3851–3950	0
2351–2450	11	3951–4050	0
2451–2550	3	4051–4150	1
2551–2650	4		

Try to find an appropriate distribution to fit these data.

(10) Length of marriage (years) was recorded for all divorces in Liège, Belgium, in 1984 (Lindsey, 1992, pp. 14–15):

Years	Divorces	Years	Divorces	Years	Divorces
1	3	19	43	37	9
2	18	20	41	38	10
3	59	21	28	39	5
4	87	22	24	40	3
5	82	23	39	41	3
6	90	24	34	42	4
7	91	25	14	43	6
8	109	26	22	44	0
9	94	27	14	45	0
10	83	28	17	46	1
11	101	29	12	47	0
12	91	30	17	48	2
13	94	31	10	49	0
14	63	32	11	50	0
15	68	33	13	51	0
16	56	34	7	52	1
17	62	35	9		
18	40	36	9		

Find a reasonable distribution to fit these data.

(11) The following table gives the number of years since their first degree of a sample of female mathematics graduates practising mathematics

or statistics (Zelterman, 1987).

Years	Number	Years	Number	Years	Number
0	5	10	2	22–23	3
1	14	11	3	24–25	4
2	10	12	3	26–27	3
3	8	13	3	28–29	1
4	11	14	0	30–31	1
5	4	15	1	32–33	2
6	3	16	5	34–35	0
7	5	17	2	36–40	6
8	7	18–19	9		
9	5	20–21	9		

Find a distribution to fit these data.

(12) The table below shows units of two types of consumer goods purchased by households over 26 weeks (Chatfield *et al.*, 1966; the frequency for 21 weeks in the last column refers to ≥ 20).

Units	Households		Units	Households	
0	1612	1498	14	0	2
1	164	81	15	0	2
2	71	47	16	0	3
3	47	25	17	2	1
4	28	16	18	0	0
5	17	17	19	0	2
6	12	6	20	1	1
7	12	10	21	0	12
8	5	3	22	2	
9	7	3	23	0	
10	6	6	24	0	
11	3	4	25	1	
12	3	4	26	2	
13	5	3	27	0	

(a) Find a suitable distribution for each good.

(b) Could the same distribution be fitted to both?

(13) The following table gives the unemployment times of laid-off personnel from the Panel Study of Income Dynamics sample (Han and Hausman, 1990). Censored individuals disappeared, so that we do not know whether they received a new job, moved, or ceased looking for employment.

Weeks	New job	Recall	Censor	Weeks	New job	Recall	Censor
1	10	93	0	36	2	1	0
2	8	118	0	37	0	1	2
3	8	55	0	38	1	0	0
4	23	58	0	39	5	4	7
5	3	18	0	40	4	1	1
6	11	26	0	41	1	0	0
7	1	6	0	42	0	0	2
8	22	38	0	43	1	4	2
9	6	13	1	44	0	0	0
10	7	10	0	45	1	0	0
11	4	4	0	46	0	0	0
12	13	32	1	47	0	0	2
13	10	19	9	48	0	0	1
14	0	9	2	49	1	0	1
15	4	14	2	50	1	1	0
16	10	9	3	51	0	0	0
17	8	7	18	52	4	0	23
18	5	2	6	53	1	0	0
19	2	0	3	54	0	0	0
20	9	12	4	55	0	0	2
21	3	1	7	56	1	0	0
22	5	7	9	57	0	0	1
23	1	0	2	58	0	0	0
24	7	10	4	59	0	0	0
25	2	1	2	60	1	0	1
26	18	15	21	61	0	0	2
27	0	2	1	62	0	0	0
28	0	2	0	63	0	0	0
29	1	0	1	64	0	0	0
30	9	4	9	65	0	0	1
31	0	0	3	66	1	0	1
32	1	0	1	67	0	1	1
33	1	0	0	68	0	0	0
34	2	1	3	69	0	1	0
35	2	0	8	70	4	3	33

(a) Find probability distributions to fit these data. Consider the censored individuals as a separate category.

(b) Are the distributions the same for new jobs, recalls, and censoring?

(14) In a case-control study of oesophageal cancer among Singapore Chinese males, the number of beverages reported drunk at burning hot

temperature was recorded (Breslow,1982):

	Controls			
Cases	0	1	2	3
0	31	5	5	0
1	12	1	0	0
2	14	1	2	1
3	6	1	1	0

Find an appropriate bivariate distribution for these data.

(15) The following table shows the numbers of lambs born to ewes in two years (Plackett, 1965, from Tallis).

	1953		
1952	0	1	2
0	58	26	8
1	52	58	12
2	1	3	9

Try to find an appropriate bivariate distribution to describe these data.

(16) Numbers of injuries and fatalities in accidents were recorded in Eastern Virginia, U.S.A., between 1 January 1969 and 31 October 1970 (Leiter and Hamdan, 1973):

	Fatalities		
Injuries	0	1	2
0	286	0	0
1	198	17	1
2	82	10	0
3	24	5	1
4	13	1	0
5	1	0	0

Look for an appropriate bivariate distribution to describe these data.

(17) The number of children ever born to a sample of mothers over 40 years of age was collected by the East African Medical Survey in the Kwimba district of Tanganyika (Brass, 1959):

Children	1	2	3	4	5	6	7	8	9	10	11	12
Mothers	49	56	73	41	43	23	18	18	7	7	3	2.

Fit a suitable truncated distribution for these data.

(18) The female knapweed gall-fly (*Urophora jaceana*) lays its eggs in batches in the unopened flowers of the black knapweed (*Centaurea nemoralis*). The second instar larva hatches from the egg and produces a hard gall-cell. Numbers of eggs and gall-cells (at different dates) were counted in flower heads in two years (Finney and Varley,

1955):

	1935		1936	
Number	Eggs	Gall-cells	Eggs	Gall-cells
1	29	287	22	90
2	38	272	18	96
3	36	196	18	57
4	23	79	11	26
5	8	29	9	10
6	5	20	6	4
7	5	2	3	5
8	2	0	0	0
9	1	1	1	1
10	0	0	0	0
11	0	0	0	0
12	1	0	0	0

(a) Find a distribution to fit such truncated data.
(b) Is the same distribution suitable for both eggs and gall-cells?
(c) Is there any change between the two years?

(19) The number of cholera cases in each house was recorded during an epidemic in India (Dahiya and Gross, 1973):

Cases	Houses
0	168
1	32
2	16
3	6
4	1

Estimate the number of houses which registered no cases, but which were already infected.

Part II

Count data

7
Counting processes

In this part, we shall look at dependent events. These will usually be events occurring to the same individual unit, instead of to independently sampled units, as was the case for the frequencies of events in the first part. Often a sequence of events over time will be available. One way to model such data is to look at the probability, called the risk or intensity, of an event in each small interval of time. Models will describe how this changes, over time, since a previous event or since the process began, and with various explanatory variables. Another equivalent way is to look at the accumulating number of events. As its name suggests, this counting process is a random variable over time, $N(t)$, which counts the number of events which have occurred up to t. Then the intensity is just the (local average) rate at which the counting process changes over time, so that the two approaches are very closely related.

7.1 Poisson processes

Often, the counts of one type of event in each fixed interval of discrete time are recorded over some period. Interest lies in the rate, or intensity, with which the events are occurring and whether this is changing over time. In an ordinary Poisson process, the intensity is constant for a given unit. If it changes, with explanatory variables or simply as a function of time, the process is said to be non-homogeneous. Such processes can easily be modelled as log linear models.

Consider perhaps the most classical example of a Poisson process, the number of deaths by horse kicks in the Prussian army from 1875 to 1894, reproduced in Table 7.1. Corps G, I, VI, and XI are noted as having a different organization than the others. The cumulative numbers of deaths over the 20 years are plotted for each corps in Figure 7.1. From the diverging step functions, we see that there is a considerable difference in the rate at which the deaths occur. The total number of deaths over the 20 years ranges from 7 to 25, with two corps having very high rates.

A model with the same homogeneous Poisson process for all corps contains just a common mean for all counts. Note that, although Table 7.1 has the usual form of a contingency table, we shall not fit the marginal

Table 7.1. Number of deaths by horse kicks in the Prussian army from 1875 to 1894 for 14 corps. (Andrews and Herzberg, 1985, p. 18)

Corps	
G	0 2 2 1 0 0 1 1 0 3 0 2 1 0 0 1 0 1 0 1
I	0 0 0 2 0 3 0 2 0 0 0 1 1 1 0 2 0 3 1 0
II	0 0 0 2 0 2 0 0 1 1 0 0 2 1 1 0 0 2 0 0
III	0 0 0 1 1 1 2 0 2 0 0 0 1 0 1 2 1 0 0 0
IV	0 1 0 1 1 1 1 0 0 0 0 1 0 0 0 0 1 1 0 0
V	0 0 0 0 2 1 0 0 1 0 0 1 0 1 1 1 1 1 1 0
VI	0 0 1 0 2 0 0 1 2 0 1 1 3 1 1 1 0 3 0 0
VII	1 0 1 0 0 0 1 0 1 1 0 0 2 0 0 2 1 0 2 0
VIII	1 0 0 0 1 0 0 1 0 0 0 0 1 0 0 0 1 1 0 1
IX	0 0 0 0 0 2 1 1 1 0 2 1 1 0 1 2 0 1 0 0
X	0 0 1 1 0 1 0 2 0 2 0 0 0 0 2 1 3 0 1 1
XI	0 0 0 0 2 4 0 1 3 0 1 1 1 1 2 1 3 1 3 1
XIV	1 1 2 1 1 3 0 4 0 1 0 3 2 1 0 2 1 1 0 0
XV	0 1 0 0 0 0 0 1 0 1 1 0 0 0 2 2 0 0 0 0

totals, as we did in the first part. There, the Poisson distribution was used to represent the multinomial distribution, as we saw in Section 1.1. In contrast, in this chapter, each individual count is assumed to have a Poisson distribution. Such a count arises from (possibly dependent) events on the same individual unit, as distinct from the frequencies of events from different units, studied up until now.

The deviance for this model is 323.23 (325.23) with 279 d.f., while that for a different intensity for each corps, fitted by introducing a factor variable, `CORPS`, reduces this by 26.14 (AIC 325.09) with 13 d.f. However, a glance at the parameter estimates for this model, in Table 7.2, reveals that the four corps with a different organization have similar, higher, rates of deaths than most of the others, although Corps XI and XIV (the latter not being one of the four with different organization) distinguish themselves by being the highest. Reducing the model to only two different intensities, for the two types of organization, say `CORPS2`, raises the deviance by 18.47 (AIC 319.56) with an increase of 12 d.f. Thus, the 1 d.f. for different organization accounts for a considerable change in deviance of 7.67.

We may also allow the process to vary with the year, a non-homogeneous Poisson process, the same within each of the two groups of corps (i.e. no interaction):

`CORPS2 + YEAR`

This reduces the deviance by a further 38.50 (AIC 319.06) with 19 d.f. However, a look at the parameters (not shown) reveals that the rate was lower in the first three years, as well as a scattering of other years. If we

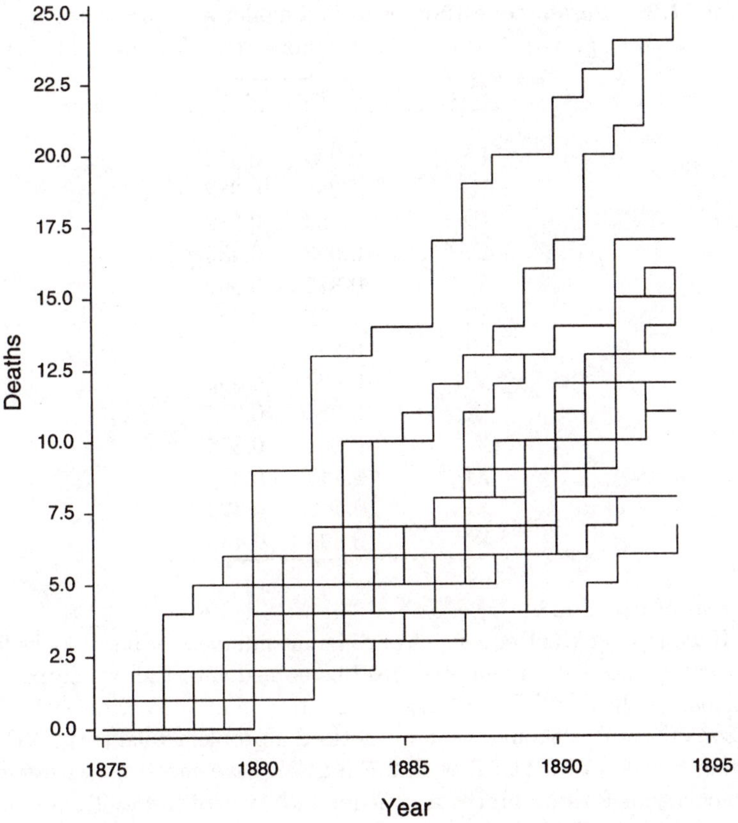

Fig. 7.1. Cumulative numbers of deaths for each Prussian army corps for the horse kick data in Table 7.1.

only allow non-homogeneity in the form of a different rate before and after this time, using a new two-level factor variable, YEARB, the model is

$$\texttt{CORPS2} + \texttt{YEARB}$$

and the deviance is raised by 28.68 (AIC 311.74) with 18 d.f., so that the break between 1877 and 1878, with one d.f., accounts for 9.82 of the change in deviance. The parameter estimates are −0.751 for the intercept, −0.416 for CORPS2, and 0.756 for the break in time, YEARB. A model with a linear trend,

$$\texttt{CORPS2} + \texttt{YEARL}$$

instead of this break fits much more poorly, with the same number of

Table 7.2. Parameter estimates from a model with differences among all corps, with respect to Corps G, for the horse kick data of Table 7.1.

Corps	Estimate	s.e.
G	0.000	—
I	0.000	0.354
II	−0.288	0.382
III	−0.288	0.382
IV	−0.693	0.433
V	−0.375	0.392
VI	0.061	0.348
VII	−0.288	0.382
VIII	−0.827	0.453
IX	−0.208	0.373
X	−0.065	0.359
XI	0.446	0.320
XIV	0.406	0.323
XV	−0.693	0.433

degrees of freedom.

If we look at the frequencies of different numbers of horse kicks for the four categories, before and after 1877 combined with the two corps organizations, we thus find that a Poisson distribution may describe each of them reasonably well, although we have noticed a problem with corps XIV. The deviance is 305.74 (311.74) with 277 d.f. We have ended up with two non-homogeneous Poisson processes, one for each type of corps. The intensity is estimated as 0.47 deaths/year for the four corps with different organization for the three years up to 1877. This is multiplied by a factor of 0.66 for the other corps (for all years) and by 2.13 after 1877 (for both corps). The difference between the two corps' organizations would obviously be considerably larger if Corps XIV were omitted from the model, but we have no reason for doing this.

If necessary, more complex relationships can be introduced between the intensity and time. However, the amount of detail possible is limited because more exact information is not available in such data about the times at which the events occurred within an observation interval.

7.2 Rates

In certain cases, the intensity of events must be weighed by the context in which they are occurring. Thus, in Section 1.4, we looked at suicide rates per day, although the data were monthly. There, we treated the data as frequencies, but they can also be thought of as counts. As we hinted there, the procedure which we used can be applied very generally. There,

Table 7.3. Lung cancer cases in four Danish cities between 1968 and 1971. (Andersen, 1977)

	City							
	Fredericia		Horsens		Kolding		Vejle	
Age	Cases	Pop.	Cases	Pop.	Cases	Pop.	Cases	Pop.
40–54	11	3059	13	2879	4	3142	5	2520
55–59	11	800	6	1083	8	1050	7	878
60–64	11	710	15	923	7	895	10	839
65–69	10	581	10	834	11	702	14	631
70–74	11	509	12	634	9	535	8	539
>75	10	605	2	782	12	659	7	619

the denominator in the rate was time. Another possibility is that it is the population size. Such models are especially important in epidemiology, in the study of the incidence or prevalence of some illness.

Let us look at the numbers of lung cancer cases, as reported in four Danish cities between 1968 and 1971, given in Table 7.3. The observed cancer rates are plotted in Figure 7.2. We are interested in determining if the lung cancer rate varies with age, and if it differs among the cities. The additive model,

$$\texttt{CITY} + \texttt{AGE}$$

with the logarithm of population as offset, has a deviance of 23.45 (41.45) with 15 d.f. We might consider simplifying the model by looking at a linear trend in age, using the centres of the age classes,

$$\texttt{CITY} + \texttt{AGEL}$$

but the deviance rises to 50.27 (60.27) with 19 d.f. However, in the rates plotted in Figure 7.2 (and in the original estimates for the factor variable age, not shown), we may have noticed a decline at the highest ages, so that a quadratic model,

$$\texttt{CITY} + \texttt{AGEL} + \texttt{AGEQ}$$

might be better. And, indeed, the deviance is 25.35 (37.35) with 18 d.f. We can also see that Horsens, Kolding, and Vejle have similar parameter estimates (not shown), so that we can contrast this with Fredericia in the model

$$\texttt{CITY2} + \texttt{AGEL} + \texttt{AGEQ} \tag{7.1}$$

which has a deviance of 25.61 (33.61) with 20 d.f. This final model, plotted

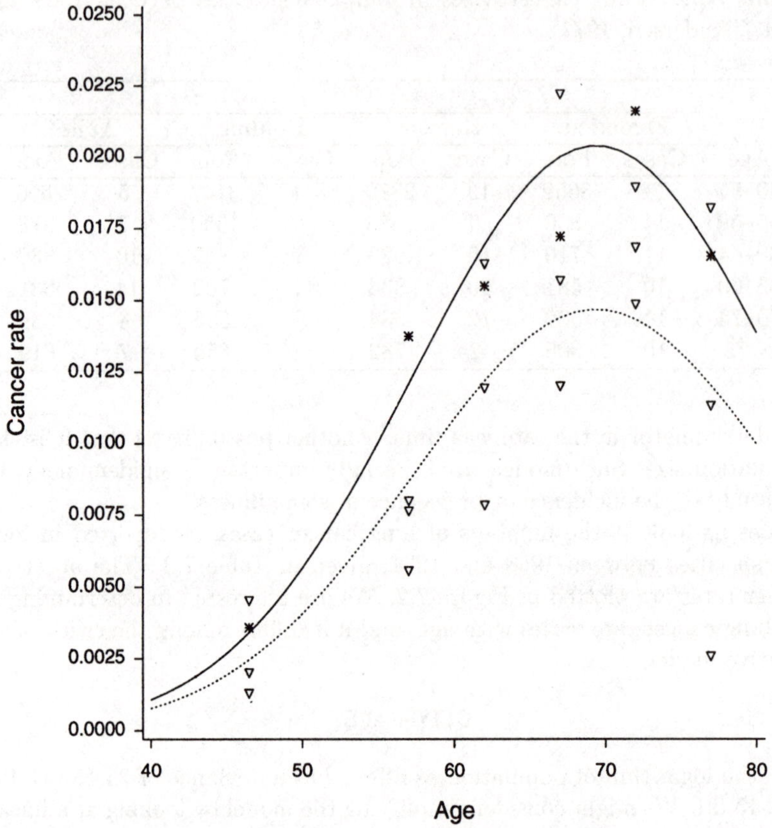

Fig. 7.2. Cancer rates with age from the model of Equation (7.1) for Fredericia (solid line and stars) and the other Danish cities (dotted line and triangles) for the data of Table 7.3.

in Figure 7.2, is

$$\begin{aligned}\log(\nu/z) &= -20.34 + 0.474x - 0.003\,42x^2 && \text{for Fredericia}\\ &= -20.67 + 0.474x - 0.003\,42x^2 && \text{for the other cities}\end{aligned}$$

with x the age and z the population. The parameter estimates show that Fredericia has a higher rate than the other cities, as can be seen in the graph.

7.3 Point processes

If the counts in a (non-homogeneous) Poisson process can be disaggregated so that the point in time (to the precision of the recording instrument) of

Table 7.4. June days with measurable precipitation (1) at Madison, Wisconsin, 1961–1971. (Klotz, 1973)

Year	
1961	10000 01101 01100 00010 01010 00000
1962	00110 00101 10000 01100 01000 00000
1963	00001 01110 00100 00010 00000 11000
1964	01000 00000 11011 01000 11000 00000
1965	10001 10000 00000 00001 01100 01000
1966	01100 11010 11001 00001 00000 11100
1967	00000 11011 11101 11010 00010 00110
1968	10000 00011 10011 00100 10111 11011
1969	11010 11000 11000 01100 00001 11010
1970	11000 00000 01000 11001 00000 10000
1971	10000 01000 10000 00111 01010 00000

each event is known, we have a point process. The data consist of a series of zeros and ones.

Consider successive June days in Madison, Wisconsin, U.S.A., with (coded 1) and without (coded 0) rain, given in Table 7.4. Here, we take June in Madison as the unit of observation, with repetitions for 11 years, but other similar data might compare different cities. The cumulative numbers of days of rain for each year are plotted in Figure 7.3. Here, there is not as much difference among the slopes as for the horse kick data in Figure 7.1. The total number of rainy days in the month ranges from seven to 15.

Because there are only two alternatives on each day, we can use a binomial distribution. If we fit a constant probability (null) model to these data, we obtain a deviance of 418.70 (420.70) with 329 d.f. However, in this special case of binary data, as we have seen, this deviance has no interpretable meaning. Let us, then, fit a model for differences among the 11 years using a factor variable, `YEAR`. The change in deviance is 13.20 (AIC 427.50) with 10 d.f., showing little indication of difference among years. The parameter estimates are given in Table 7.5. We note that there may be some cyclical effect, with higher precipitation in the years 1966 to 1969. This could be investigated using the methods of Section 1.4. Other trends in time could also be looked at; if the series were longer within each year, seasonal fluctuations might be considered.

Suppose, now, that, as is the case here, only one event could occur in a given interval and that, at a given time point, it depends only on the event at the immediately preceding point. This is the hypothesis of a first-order Markov chain and is the simplest example of one. We have a square 2×2 transition matrix of conditional probabilities, $\mathbf{T}_t$, of the events, which may vary in time. If the rows are the states at the previous time point and the

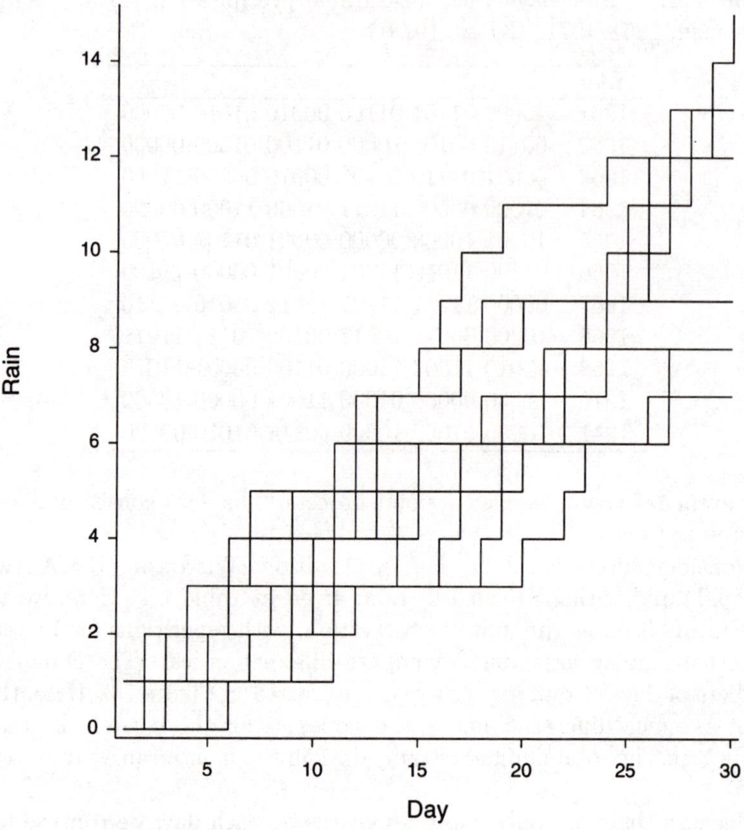

Fig. 7.3. Cumulative numbers of days of rain for each year, from Table 7.4.

columns are the present states, then the row probabilities sum to one. Premultiplying the column vector, $\mathbf{n}_t$, of frequencies of units in the different states (the marginal frequencies) at a given time point, t, by the transpose of this matrix will give the vector for the next time period, $t+1$:

$$\mathbf{n}_{t+1} = \mathbf{T}_t^T \mathbf{n}_t \tag{7.2}$$

Thus, the transition matrix represents the pattern of change. The estimated unconditional probability of precipitation, the number of ones in the table divided by the total number of days, is 0.32, while the probability of rain, given rain the previous day, is 0.41 and, given dry the previous day, is 0.28. Thus, the transition matrix, if assumed constant over the months and years, is

Table 7.5. Parameter estimates from a model with differences among years, with respect to 1961, for the rainfall data of Table 7.4.

Year	Estimate	s.e.
1961	0.000	—
1962	−0.164	0.573
1963	−0.164	0.573
1964	−0.164	0.573
1965	−0.342	0.587
1966	0.442	0.545
1967	0.714	0.541
1968	0.847	0.540
1969	0.579	0.542
1970	−0.342	0.587
1971	−0.164	0.573

$$\mathbf{T} = \begin{pmatrix} 0.72 & 0.28 \\ 0.59 & 0.41 \end{pmatrix}$$

These probabilities can be obtained from a simple binary logistic model, applied directly to the raw data, as in Section 3.4, with a zero/one indicator of precipitation and a lagged variable for what happened the previous day (`LAG`). The difference in deviance between models with and without the lagged variable is 5.25 (AICs 64.84 and 61.69) with 1 d.f., while adding a factor variable for differences among years,

$$\texttt{LAG} + \texttt{YEAR}$$

further reduces it by 11.82 (AIC 69.76) with 10 d.f. The latter confirms that the transition matrix seems to be constant over years. Introduction of lagged variables further back in time changes the deviance very little.

Because there are no time-varying explanatory variables, the same results could also be obtained by classifying the 330 observations in Table 7.4 according to rain on a given day, rain the previous day, and year, to give a $2 \times 2 \times 11$ contingency table of counts, and applying a log linear or logistic model. Of course, higher dimensional tables would be required to look at lags further back in time.

7.4 Multiplicative intensity models

So far in this chapter, we have studied stochastic process models in discrete time. We shall now look at some continuous-time models, although any observations will actually be in discrete time, limited by the precision of the measuring instrument.

An event history follows a unit over time, recording the occurrence of events. These may either be repetitions of the same event, as in the previous sections,. or different events. An event may signal a change of state of a unit. Thus, we must allow the intensity function to vary over time and model it directly as a non-homogeneous Poisson process.

In fact, except for the exponential distribution, the intensity function does, in any case, change over time for any duration distribution which might be used for such data, but only as a strict function of time. Thus, for example, for the Weibull distribution, the intensity function is

$$\lambda(t; \boldsymbol{\beta}, \kappa) = t^{\kappa-1} h(\boldsymbol{\beta}, \mathbf{x})$$

where $h(\cdot)$ is some function of the explanatory variables. This is a member of the proportional hazards or multiplicative intensities family. Often, we shall also want to introduce observed explanatory variables which may change over time, even in between events.

To go further, it is fruitful to consider the counting process approach in more detail. Let the intensity of the process be $\lambda(t|\mathcal{F}_{t-}; \boldsymbol{\beta})$ such that

$$\lambda(t|\mathcal{F}_{t-}; \boldsymbol{\beta})dt = \Pr[dN(t) = 1|\mathcal{F}_{t-}; \boldsymbol{\beta}]$$

where $\boldsymbol{\beta}$ is a vector of unknown parameters and $\mathcal{F}_{t-}$ is the filtration, or history, up to, but not including, t. Then, the kernel of the log likelihood function for observation over the continuous time interval, $(0, T]$ is (Borgan, 1984, Andersen and Borgan, 1985)

$$\log[\mathrm{L}(\boldsymbol{\beta})] = \int_0^T \log[\lambda(t|\mathcal{F}_{t-}; \boldsymbol{\beta})]dN(t) - \int_0^T \lambda(t|\mathcal{F}_{t-}; \boldsymbol{\beta})I(t)dt$$

where $I(t)$ is an indicator variable, with value one if the process is under observation at time t and zero otherwise.

Now, in any empirical situation, the process will only be observed at discrete time intervals, once an hour, once a day, once a week. Suppose that these are sufficiently small so that at most one event occurs in any interval, although there will be a finite non-zero probability of more than one. With M intervals of observation, not all necessarily the same size, and using the same approximation to the integrals as in Chapter 6, the (approximate) log likelihood becomes

$$\log[\mathrm{L}(\boldsymbol{\beta})] = \sum_{t=1}^{M} \log[\lambda(t|\mathcal{F}_{t-}; \boldsymbol{\beta})]\Delta N(t) - \sum_{t=1}^{M} \lambda(t|\mathcal{F}_{t-}; \boldsymbol{\beta})I(t)\Delta_t$$

where Δ_t is the width of the tth observation interval and $\Delta N(t)$ is the change in the count during that interval, almost always with values zero or one. This is just the kernel of the log likelihood of a (censored) Poisson distribution for $\Delta N(t)$, with mean, $\lambda(t|\mathcal{F}_{t-}; \boldsymbol{\beta})\Delta_t$. Conditional on the filtration, it is the likelihood for a (local) Poisson process. The structure which we shall place on this likelihood will determine what stochastic process we are modelling (Lindsey, 1995b).

Let us look at data on the treatment of chronic granulotomous disease from a placebo controlled randomized clinical trial of gamma interferon, analysed by Fleming and Harrington (1990, pp. 162–163, 377–383, not reproduced here). Interest centres on the time between successive infections. However, of the 128 subjects, only 44 had one or more serious infections, for a total of 76 infections. A number of variables describing each patient are also available: 1. type of inheritance, 2. age, 3. height, 4. weight, 5. use of corticosteroids at entry, 6. use of antibiotics at entry, 7. sex, and 8. location of the hospital.

Observations were recorded each day. The Poisson response variable will be a vector as long as the total number of observation days for all patients together (37 477 days). It will contain zeros except for the 76 days on which an infection occurred, as indicated by ones. The `TIME` variable contains successive integers counting the number of days until an infection occurs, after which it is reinitialized to one. The `TREATMENT` variable contains one for all days for all patients under gamma interferon treatment and two for those under placebo. The explanatory variables, such as age, contain the same value for all days for a given patient. (Because the relative ages of patients contain the pertinent information, ages do not change with time in the model.)

We shall fit the exponential, extreme value, Weibull, and Cox models; these have, respectively, no time variable, linear time, log time, and a factor variable for time from the previous event. Thus, for example, a Weibull model for treatment effect would be

$$\texttt{LOGTIME} + \texttt{TREATMENT}$$

and a Cox model for age would be

$$\texttt{TIME} + \texttt{TREATMENT}$$

where `TIME` is a factor variable.

A few of the relevant deviances are given in the upper panel of Table 7.6. The simple exponential model fits about as well as the extreme value and Weibull models, but the Cox model is considerably better, although it does have 74 extra parameters. A treatment effect is plausible and age marginally so; no other variable changes the deviance very much at all. The log intensity is about 1.09 higher for the placebo group as compared to treatment, while it decreases by about 0.027 for each year of increase in age.

The reader should notice that, in our models, we have treated all infections for a given patient identically. No dependence structure has been

Table 7.6. Deviances (AIC) of models for the data on the treatment of chronic granulotomous disease for all patients.

	Exp.	Extreme Value	Weibull	Cox
Null	942.5 (944.5)	942.2 (946.2)	939.9 (943.9)	707.5 (861.5)
Treat	923.9 (927.9)	923.9 (929.9)	922.5 (928.5)	688.6 (846.6)
Age	937.9 (941.9)	937.7 (943.7)	935.5 (941.5)	703.0 (859.0)
		Birth Process		
Null	915.1 (919.1)	913.6 (919.6)	915.1 (921.1)	681.3 (837.3)
Treat	905.4 (911.4)	903.4 (911.4)	905.4 (913.4)	670.5 (828.6)
Inter	904.5 (912.5)	902.4 (912.4)	904.5 (914.5)	669.3 (829.3)
Age	912.3 (918.3)	910.7 (918.7)	912.3 (920.3)	678.5 (836.5)

introduced to link recurrent infections.

7.5 Birth processes

The probability of an event may depend on previous history, the filtration, in many ways. One useful model is to have it depend on the number of previous events, N_t, called a birth model. This can simply be introduced into the log linear model of the counting process described in the previous section.

Let us now fit such a birth process to the data on chronic granulotomous disease, taking into account the number of infections which have already occurred. Because N_t is zero in the first period, we introduce a term, $\nu \log(N_t + 1)$ (LOGN), into the log linear model. In this way, we obtain a form of generalized birth model:

$$\lambda(t; \boldsymbol{\beta}, \kappa) = (N_t + 1)^{\nu} \lambda_0(t, \kappa) h(\boldsymbol{\beta})$$

where $\lambda_0(t, \kappa)$ is the appropriate function for the distribution concerned (e.g. $t^{\kappa-1}$ for the Weibull). Thus, the Weibull birth process will be fitted by

$$\texttt{LOGN} + \texttt{LOGTIME} + \texttt{TREATMENT}$$

The deviances are presented in the lower panel of Table 7.6. These models fit much better than the corresponding independence ones. The Weibull model is now even closer to the exponential.

When an interaction is introduced between treatment and the variable describing the number of events, for example,

$$\texttt{LOGN} + \texttt{LOGTIME} + \texttt{TREATMENT} + \texttt{LOGN} \cdot \texttt{TREATMENT}$$

for the Weibull model, the fit is not changed. The difference in log intensity for treatment is 0.87 (for the Cox model, with s.e. 0.27) before the first event. The estimate of the regression coefficient for the event variable is 1.00 (s.e. 0.23). The intensity is higher under placebo, but increases at each infection in the same way for treatment and placebo.

The models which we have been studying are in the family of multiplicative intensities. A second family, of additive models, has been proposed by Aalen (1989). These models may be fitted in exactly the same way as for the previous ones; the log link is simply changed to the identity.

For this data set, the exponential model, with treatment and birth effects, has a deviance of 900.11 (906.11), considerably better than the parametric multiplicative intensity models in the table above. Removing the number of infections (the birth effect) increases the deviance by 23.33 (927.44). The Weibull and Cox models give negative intensities, indicating inappropriate models.

In conclusion, we have demonstrated that the independence assumption, among successive infections on the same patient, used by Fleming and Harrington (1990), is questionable, as, perhaps, is the multiplicative intensities assumption.

7.6 Learning models

Learning models are a generalization of birth models. In the latter, the probability of an event is made to depend only on the number of previous events. In a learning model, it can also depend on the number of non-events.

Consider the Solomon–Wynne experiment on dogs, whereby they learn to avoid a shock (Kalbfleisch, 1985, pp. 83–88). A dog is in a compartment with a floor through which a shock can be applied. The lights are turned out and a barrier raised; ten seconds later, the shock occurs. Thus, the dog has ten seconds, after the lights go out, to jump the barrier and avoid the shock. Each of 30 dogs is subjected to 25 such trials. The results, in Table 7.7, are recorded as a shock trial ($y_{it} = 0$), when dog i remains in the compartment at time t, or an avoidance trial ($y_{it} = 1$), when it jumps out before the shock. The cumulative numbers of avoidances for the first 15 dogs are plotted in Figure 7.4. Here, we see that the dogs are very similar in their learning behaviour, at least after the first few trials.

Our model supposes that a dog learns from previous trials. The probability of avoidance may depend on the number of previous shocks (`SHOCK`) and on the number of previous avoidances (`AVOID`). Let π_{it} be the probability of a shock to dog i at trial t ($t = 0, \ldots, 24$), given its reactions on previous trials, and N_{it} be the number of avoidances before trial t. Then $M_{it} = t - N_{it}$ is the number of previous shocks. We use the model

$$\pi_{it} = \kappa^{N_{it}} \upsilon^{M_{it}} \tag{7.3}$$

Table 7.7. The Solomon–Wynne dog experiment with each line giving the results of 25 trials for one dog. (Kalbfleisch, 1985, pp. 83–88)

00101	01111	11111	11111	11111
00000	00100	00001	11111	11111
00000	11011	00110	10111	11111
01100	11110	10101	11111	11111
00000	00011	11111	11111	11111
00000	01111	00101	11111	11111
00000	10000	00111	11111	11111
00000	00110	01111	11111	11111
00000	10101	10100	01111	10110
00001	00110	10111	11111	11111
00000	00000	11111	10111	11111
00000	11111	00111	11111	11111
00011	01001	11111	11111	11111
00001	01101	11111	11111	11111
00010	11011	11111	11111	11111
00000	00111	11111	11111	11111
01010	00101	11101	11111	11111
00001	01011	11101	11111	11111
01000	01000	11111	11111	11111
00001	10101	10101	11111	11111
00011	11101	11111	11111	11111
00101	01111	11111	10011	11111
00000	00111	11111	11111	11111
00000	00011	10100	01101	11111
00000	01011	11010	11111	11111
00101	11011	01111	11111	11111
00001	01111	11111	11111	11111
00010	10111	01011	11111	11111
00001	10011	10101	01011	11111
00001	11111	01011	11111	11111

or

$$\log(\pi_{it}) = \alpha N_{it} + \beta M_{it} \tag{7.4}$$

where $\alpha = \log(\kappa)$ and $\beta = \log(\upsilon)$.

Because

$$\frac{\pi_{it}}{\pi_{i,t-1}} = \begin{cases} \kappa \text{ if } y_{i,t-1} = 1 \\ \upsilon \text{ if } y_{i,t-1} = 0 \end{cases} \tag{7.5}$$

we see that the probability of shock changes by a factor of κ if there was

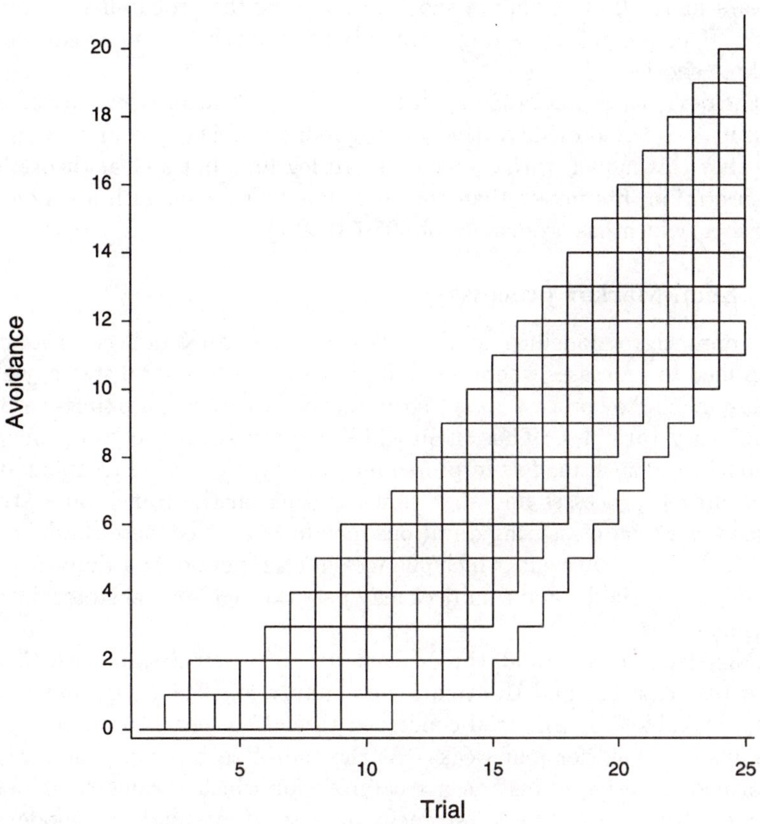

Fig. 7.4. Cumulative numbers of avoidances in the learning experiment for the first 15 dogs in Table 7.7.

an avoidance at the previous trial or υ if there was a shock.

In contrast to the counting processes of the previous sections, where there was a small probability of more than one event in an observation interval, here, only one event, either a shock or an avoidance, can occur. Thus, as in Section 7.3, we shall use a binomial distribution. Another context in which this would be appropriate occurs when the event is absorbing, as in mortality studies: death can only happen once.

We fit the model

$$\texttt{SHOCK} + \texttt{AVOID}$$

with a logindexlog link link. The parameters are estimated as $\hat{\alpha} = -0.237$ (s.e. 0.0237) and $\hat{\beta} = -0.0791$ (s.e. 0.0119). Hence, $\hat{\kappa} = \exp(-0.237) = 0.789$ and $\hat{\upsilon} = \exp(-0.0791) = 0.924$, indicating that an avoidance trial appears more effective than a shock in reducing the probability of future shock. Because $\hat{\kappa} \cong \hat{\upsilon}^3$, a dog learns about as much by one avoidance as by three shocks.

The deviance is 552.2 (556.2) with 718 d.f., but, because these are binary data, it does not provide a measure of goodness of fit. It is interesting to note that this model, with a non-standard log link, but a clear theoretical interpretation, fits better than the corresponding model with a canonical logit link, which has a deviance of 605.7 (609.7).

7.7 Semi-Markov processes

The approach to modelling counting processes described in Section 7.4 can be applied to processes where the unit passes between several states, yielding a semi-Markov process. It is also applicable when there are time-varying explanatory variables. Thus, it provides a synthesis of the two principal approaches to such models in probability theory, that using sample paths and counting processes and that concentrating on the transition matrix. A cross-over trial measuring durations may have both of these characteristics. Thus, the treatment, which changes in each period, is a time-varying explanatory variable, and a subject may pass through several states during treatment.

Consider a three-period, three-treatment cross-over design, with 15 patients participating and the treatments simply labelled A, B, and C, as shown in Table 7.8. The trial concerned three treatments for angina pectoris, each applied for four weeks. At the end of each period, the patient performed an exercise test on a treadmill, for which various times were recorded: time to 1 mm ST segment depression, to mild and moderate pain, and total time of exercise, as decided by the doctor, stopping being due to excessive anginal pain, tiredness, breathlessness, or other symptoms. Only the end of the exercise was always uncensored. The total time under observation for all subjects was 1908 ten-second periods, which will be the length of the vectors.

The latter three events mentioned had to occur in the order given, but ST segment depression could occur at any time (or not at all) in the exercise period. Also, in this particular study, moderate pain, when it was recorded, always occurred simultaneously with the end of the exercise, apparently being the cause of stopping in those cases. When it did not occur (was censored), stopping must have been for other reasons. Thus, the time from moderate pain to stopping was always zero, the former either censored or not and the latter never censored. Hence, we shall use, as the two responses, time from the beginning of the exercise to mild pain, and from mild to moderate pain, if the former is not censored. Total times from the beginning of the exercise to each event are given in Table 7.8, including time to 1 mm ST segment depression.

Table 7.8. Times (ten-second intervals) to mild and moderate pain and to 1 mm ST segment depression, with censored values indicated by an asterisk, in an exercise cross-over trial. (Lindsey *et al.*, 1995)

Id	Treat.	Mild pain	Mod. pain	ST dep.	Treat.	Mild pain	Mod. pain	ST dep.
		Period 1				Period 2		
1	A	36	42	36	B	36	36*	30
2	B	36*	36*	36	C	48	48*	48
3	C	54*	54*	54	A	48	48*	48
4	A	24	36	24	C	36*	36*	36
5	C	63*	63*	63	B	48	60	36
6	B	54	60	48	A	36	36*	36
7	B	27	27*	24	A	36	36*	36
8	A	36*	36*	36	C	24*	24*	24
9	C	66	66*	60	B	63	63*	60
11	A	48	48*	48*	B	60*	60*	48
12	C	24	24*	24	A	24	24*	24
13	C	72*	72*	72	A	60*	60*	60
14	A	24	36	24	B	24	36	12
17	A	24	27	24	C	36	42	36
18	C	24	36	24	B	36	36*	24*
		Period 3						
1	C	36*	36*	36				
2	A	36	48	36				
3	B	36	36*	36				
4	B	36*	36*	36				
5	A	24	36	24				
6	C	24*	24*	24				
7	C	36	36*	36				
8	B	36*	36*	36				
9	A	48	60	60				
11	C	48*	48*	36				
12	B	12	24	24				
13	B	72	72*	60				
14	C	36	36*	36				
17	B	24	36	24				
18	A	24	36	24				

As previously, several parametric distributions, all with multiplicative intensity functions, will provide the basis of our models: the exponential, Weibull, and extreme value. 1 mm ST segment depression will be a time-varying explanatory variable, with value zero before it occurs and one after. We also have the explanatory variables, treatment, period, and carryover

Table 7.9. Changes in deviances for adding or removing a given effect or adding an interaction, with respect to an extreme value model containing pain type and treatment, which has a deviance of 173.51 (211.51), for the three-period cross-over trial of Table 7.8. Interactions, with d.f., are with respect to pain type.

	Remove			Add		
	Extreme value	Pain type	Treatment	St Dep	Period	Carry over
Effect	88.25	12.72	23.57	1.71	4.76	5.17
AIC	297.76	222.23	231.08	211.80	210.75	212.34
Interaction	1.93	—	0.22	1.25	3.62	3.75
AIC	211.58	—	215.29	212.55	211.14	214.59
d.f.	1	1	2	1	2	3

effect. Stratification on subjects will be used. With two times observed (those to mild and moderate pain), the intensity function might differ between them. This is allowed for in the log linear model by an indicator variable. Such differences could occur in several basically different ways.

- The underlying distribution might be different, for example, Weibull in one and extreme value in the other. Then, in this example, both time and log time would interact with the pain-type indicator variable, a different one of the two cancelling out in each of the two periods.
- In a simpler case, the same distribution would be suitable for both times, but with a different parameter value. This would be indicated by an interaction between the indicator variable and the appropriate time variable.
- The explanatory variables might influence the intensity function differently for the two pain types. Here, we would find an interaction between the pain-type indicator and that explanatory variable.

Thus, a variety of fundamentally different models can easily be fitted using the log linear model methodology.

When these models are applied to the data, we discover that the extreme value distribution gives a considerably superior fit over the two others. The level of the intensity function also varies significantly between the two pain-type times and among periods. Finally, there is a treatment effect. However, no interactions with pain-type seem necessary. The results for our final model,

$$\texttt{TIME} + \texttt{PAINTYPE} + \texttt{TREATMENT} + \texttt{PERIOD} + \texttt{PATIENT}$$

with deviance 168.75 (210.75) are summarized in Table 7.9.

The parameter estimates are 0.220 (s.e. 0.033) for the extreme value power parameter, 2.543 (s.e. 0.720) for the difference in log intensity of time from mild to moderate pain as compared to time from start to mild pain, −0.474 (s.e. 0.578) and 0.644 (s.e. 0.543) for the differences in periods 2 and 3 as compared to 1, and −1.082 (s.e. 0.509) and −2.938 (s.e. 0.643) for the differences in treatments B and C with respect to A. Thus, the intensity function for mild pain is lower than that for moderate pain, indicating that the time from starting exercise to mild pain is usually longer than that from mild to moderate pain. In addition, C is the superior treatment, and A inferior, for reducing intensity (risk), i.e. increasing time both to mild and moderate pain. For some reason, the intensity of events is lower in the second period than in the other two.

Time to ST segment depression becomes necessary in the model if treatment is removed from it. The parameter estimate is then 0.956 (s.e. 0.498). Once this event has occurred, the risk of pain increases, although this is hidden by the difference in risk with treatment.

For the same two explanatory variables, pain type and treatment, the deviance is 3.54 larger with the Weibull distribution, as compared to extreme value, both with the same number of parameters. If we consider the same extreme value model, but without stratification on subjects, the deviance rises by 80.96 (14 d.f.) and the treatment effects are reduced to about one-half the size given above.

In this example of a semi-Markov process, each subject passes in order through the states, from mild pain, perhaps to moderate pain, and then stopping the exercise. If the units can pass in any direction between several states, the polytomous regression techniques of Section 3.5 must be used, which means that the vectors in the model will be even longer.

7.8 Capture–recapture models

One special counting process, which is much used in animal ecology, is the capture–recapture model for estimating populations. This is closely related to the population estimation techniques of Section 5.4, but here it refers to a dynamic process in time. Animals are captured, given some distinctive identifying mark which is recorded, and released. This procedure is repeated several times, but, on each subsequent time, only previously uncaptured animals are marked, while recaptures of previously marked animals are also recorded. This provides a time sequence indicating whether each animal was captured at each time. As in Section 5.4, animals never captured will not be represented in the results.

One of the major studies of this kind was carried out over 25 years on the eider duck (*Somateria mollissima*). Obviously, such data would create much too large a table if all possibilities were enumerated. Table 7.10 shows the results for the years 20 to 23, but with extra categories added at the beginning and end to indicate if a duck was ever captured before and after

Table 7.10. Capture–recapture data on the eider duck (*Somateria mollissima*) for six periods. (Cormack, 1989, from Coulson)

Capture	Frequency	Capture	Frequency
YYYYYY	13	NYYYYY	4
YNYYYY	5	NNYYYY	4
YYNYYY	7	NYNYYY	3
YNNYYY	7	NNNYYY	6
YYYNYY	5	NYYNYY	2
YNYNYY	5	NNYNYY	8
YYNNYY	5	NYNNYY	2
YNNNYY	2	NNNNYY	22
YYYYNY	6	NYYYNY	0
YNYYNY	3	NNYYNY	5
YYNYNY	2	NYNYNY	2
YNNYNY	12	NNNYNY	15
YYYNNY	5	NYYNNY	2
YNYNNY	10	NNYNNY	14
YYNNNY	7	NYNNNY	6
YNNNNY	9	NNNNNY	79
YYYYYN	3	NYYYYN	0
YNYYYN	6	NNYYYN	3
YYNYYN	4	NYNYYN	0
YNNYYN	5	NNNYYN	3
YYYNYN	4	NYYNYN	0
YNYNYN	3	NNYNYN	6
YYNNYN	4	NYNNYN	2
YNNNYN	7	NNNNYN	40
YYYYNN	6	NYYYNN	0
YNYYNN	3	NNYYNN	3
YYNYNN	5	NYNYNN	1
YNNYNN	7	NNNYNN	16
YYYNNN	22	NYYNNN	11
YNYNNN	23	NNYNNN	27
YYNNNN	25	NYNNNN	18
YNNNNN	362	NNNNNN	—

that period.

A first model assumes a closed population. A zero–one variable, `Pi`, indicating whether or not the duck was captured in year i, is fitted,

$$\texttt{P1} + \texttt{P2} + \texttt{P3} + \texttt{P4} + \texttt{P5} + \texttt{P6}$$

This model has a deviance of 655.22 (669.22) with 56 d.f. A second model introduces births, or immigrants. The second period has the indicator variable, P1·P2, that for the third period, P1·P2·P3, and so on. Adding this to our model, we obtain

$$\texttt{P1} + \texttt{P2} + \texttt{P3} + \texttt{P4} + \texttt{P5} + \texttt{P6} + \texttt{P1} \cdot \texttt{P2} + \texttt{P1} \cdot \texttt{P2} \cdot \texttt{P3} + \texttt{P1} \cdot \texttt{P2} \cdot \texttt{P3} \cdot \texttt{P4} + \texttt{P1} \cdot \texttt{P2} \cdot \texttt{P3} \cdot \texttt{P4} \cdot \texttt{P5}$$

which yields a deviance of 621.01 (643.01) with 52 d.f. Next, we can add the possibility of deaths; birds may die, so that they cannot be recaptured. This time, we start at the other end: P5·P6, and continue with P4·P5·P6, and so on. The death only model is, then,

$$\texttt{P1} + \texttt{P2} + \texttt{P3} + \texttt{P4} + \texttt{P5} + \texttt{P6} + \texttt{P2} \cdot \texttt{P3} \cdot \texttt{P4} \cdot \texttt{P5} \cdot \texttt{P6} + \texttt{P3} \cdot \texttt{P4} \cdot \texttt{P5} \cdot \texttt{P6} + \texttt{P4} \cdot \texttt{P5} \cdot \texttt{P6} + \texttt{P5} \cdot \texttt{P6}$$

with a deviance of 255.38 (277.38) and 52 d.f. Reintroducing births, we have the birth and death model,

$$\texttt{P1} + \texttt{P2} + \texttt{P3} + \texttt{P4} + \texttt{P5} + \texttt{P6} + \texttt{P1} \cdot \texttt{P2} + \texttt{P1} \cdot \texttt{P2} \cdot \texttt{P3} + \texttt{P1} \cdot \texttt{P2} \cdot \texttt{P3} \cdot \texttt{P4} + \texttt{P1} \cdot \texttt{P2} \cdot \texttt{P3} \cdot \texttt{P4} \cdot \texttt{P5} + \texttt{P2} \cdot \texttt{P3} \cdot \texttt{P4} \cdot \texttt{P5} \cdot \texttt{P6} + \texttt{P3} \cdot \texttt{P4} \cdot \texttt{P5} \cdot \texttt{P6} + \texttt{P4} \cdot \texttt{P5} \cdot \texttt{P6} + \texttt{P5} \cdot \texttt{P6}$$

which has a deviance of 83.13 (113.13) and 48 d.f. However, in this model, P5·P6 has an impossible negative parameter estimate and must be eliminated, giving a deviance of 83.36 (111.36). We can now introduce trap dependence. This is the idea that some ducks are either systematically attracted or repelled by the capture methods. The variables are simply products of successive period indicators: P1·P2 and so on. The birth and death model with trap dependence,

$$\texttt{P1} + \texttt{P2} + \texttt{P3} + \texttt{P4} + \texttt{P5} + \texttt{P6} + \texttt{P1} \cdot \texttt{P2} + \texttt{P1} \cdot \texttt{P2} \cdot \texttt{P3} + \texttt{P1} \cdot \texttt{P2} \cdot \texttt{P3} \cdot \texttt{P4} + \texttt{P1} \cdot \texttt{P2} \cdot \texttt{P3} \cdot \texttt{P4} \cdot \texttt{P5} + \texttt{P2} \cdot \texttt{P3} \cdot \texttt{P4} \cdot \texttt{P5} \cdot \texttt{P6} + \texttt{P3} \cdot \texttt{P4} \cdot \texttt{P5} \cdot \texttt{P6} + \texttt{P4} \cdot \texttt{P5} \cdot \texttt{P6} + \texttt{P2} \cdot \texttt{P3} + \texttt{P3} \cdot \texttt{P4} + \texttt{P4} \cdot \texttt{P5}$$

has a deviance of 78.05 (114.05) with 45 d.f. Finally, a constant effort model replaces the central period indicators, P2 to P5 by CEFF=P2+P3+P4+P5, i.e. probability of capture is assumed constant in these periods, usually because nets were always set for the same amount of time. Thus, the constant effort birth and death model, without trap dependence,

Table 7.11. Parameter estimates from the model of Equation (7.6) for the capture–recapture data of Table 7.10.

	Estimate	s.e.
1	1.301	0.170
P1	−0.781	0.155
CEFF	0.244	0.068
P6	−0.145	0.114
P1·P2	0.978	0.172
P1·P2·P3·P4	1.272	0.171
P1·P2·P3·P4·P5	0.621	0.187
P2·P3·P4·P5·P6	2.383	0.139
P4·P5·P6	1.375	0.143

$$\begin{aligned}\mathtt{P1} + \mathtt{CEFF} + \mathtt{P6} + \mathtt{P1}\cdot\mathtt{P2} + \mathtt{P1}\cdot\mathtt{P2}\cdot\mathtt{P3} + \mathtt{P1}\cdot\mathtt{P2}\cdot\mathtt{P3}\cdot\mathtt{P4} \\ +\mathtt{P1}\cdot\mathtt{P2}\cdot\mathtt{P3}\cdot\mathtt{P4}\cdot\mathtt{P5} + \mathtt{P2}\cdot\mathtt{P3}\cdot\mathtt{P4}\cdot\mathtt{P5}\cdot\mathtt{P6} \\ +\mathtt{P3}\cdot\mathtt{P4}\cdot\mathtt{P5}\cdot\mathtt{P6} + \mathtt{P4}\cdot\mathtt{P5}\cdot\mathtt{P6}\end{aligned} \tag{7.6}$$

has a deviance of 84.78 (108.78) with 51 d.f.

The preferable model appears to be this final one. However, it may be slightly improved by removing P3·P4·P5·P6, P5·P6, and P1·P2·P3, indicating that there were no deaths or births at those moments, with a deviance of 89.21 (107.21) and 54 d.f. The parameter estimates are given in Table 7.11. Cormack (1985, 1989) gives more detailed information on fitting and interpreting such models.

7.9 Multivariate intensity functions

In a similar way to our construction of multivariate distributions in Section 6.4, we can model multivariate intensity functions. In the simplest case, a bivariate intensity without explanatory variables, we shall have a two-dimensional table, such as the data given in Table 7.12, involving two processes over time. In fact, the models for the deaths by horse kicks in the Prussian army in Table 7.1 might already be thought of as a bivariate intensity function, with one process in time and the other in space.

In the study of the incidence of acquired immune deficiency syndrome (AIDS), one major problem involves delays in reporting diagnosed cases. Cases which are diagnosed locally at a certain time may only reach the collection centre some months later. Such data on incidence of AIDS, with reporting delays, take the form of a rectangular contingency table with observations in one triangular corner missing. The two dimensions of such a table are the diagnosis period, describing the process of growing AIDS incidence, and the process causing reporting delay. The margin for

Table 7.12. Number of AIDS reports in England and Wales to the end of 1992 by quarter, with reporting delays. (de Angelis and Gilks, 1994)

	Delay period (quarters)														
Year	0	1	2	3	4	5	6	7	8	9	10	11	12	13	14+
1983	2	6	0	1	1	0	0	1	0	0	0	0	0	0	1
	2	7	1	1	1	0	0	0	0	0	0	0	0	0	0
1984	4	4	0	1	0	2	0	0	0	0	2	1	0	0	0
	0	10	0	1	1	0	0	0	1	1	1	0	0	0	0
	6	17	3	1	1	0	0	0	0	0	0	1	0	0	1
	5	22	1	5	2	1	0	2	1	0	0	0	0	0	0
1985	4	23	4	5	2	1	3	0	1	2	0	0	0	0	2
	11	11	6	1	1	5	0	1	1	1	1	0	0	0	1
	9	22	6	2	4	3	3	4	7	1	2	0	0	0	0
	2	28	8	8	5	2	2	4	3	0	1	1	0	0	1
1986	5	26	14	6	9	2	5	5	5	1	2	0	0	0	2
	7	49	17	11	4	7	5	7	3	1	2	2	0	1	4
	13	37	21	9	3	5	7	3	1	3	1	0	0	0	6
	12	53	16	21	2	7	0	7	0	0	0	0	0	1	1
1987	21	44	29	11	6	4	2	2	1	0	2	0	2	2	8
	17	74	13	13	3	5	3	1	2	2	0	0	0	3	5
	36	58	23	14	7	4	1	2	1	3	0	0	0	3	1
	28	74	23	11	8	3	3	6	2	5	4	1	1	1	3
1988	31	80	16	9	3	2	8	3	1	4	6	2	1	2	6
	26	99	27	9	8	11	3	4	6	3	5	5	1	1	3
	31	95	35	13	18	4	6	4	4	3	3	2	0	3	3
	36	77	20	26	11	3	8	4	8	7	1	0	0	2	2
1989	32	92	32	10	12	19	12	4	3	2	0	2	2	0	2
	15	92	14	27	22	21	12	5	3	0	3	3	0	1	1
	34	104	29	31	18	8	6	7	3	8	0	2	1	2	–
	38	101	34	18	9	15	6	1	2	2	2	3	2	–	–
1990	31	124	47	24	11	15	8	6	5	3	3	4	–	–	–
	32	132	36	10	9	7	6	4	4	5	0	–	–	–	–
	49	107	51	17	15	8	9	2	1	1	–	–	–	–	–
	44	153	41	16	11	6	5	7	2	–	–	–	–	–	–
1991	41	137	29	33	7	11	6	4	–	–	–	–	–	–	–
	56	124	39	14	12	7	10	–	–	–	–	–	–	–	–
	53	175	35	17	13	11	–	–	–	–	–	–	–	–	–
	63	135	24	23	12	–	–	–	–	–	–	–	–	–	–
1992	71	161	48	25	–	–	–	–	–	–	–	–	–	–	–
	95	178	39	–	–	–	–	–	–	–	–	–	–	–	–
	76	181	–	–	–	–	–	–	–	–	–	–	–	–	–
	67	–	–	–	–	–	–	–	–	–	–	–	–	–	–

diagnosis period, in which we are particularly interested, gives the total incidence over time, but with the most recent values too small because of the missing triangle of values not yet reported.

Let us write a simple model for the intensity function of a bivariate Poisson process, with the reporting delay process stationary over time, as

$$\lambda(t, u) = \lambda_{Dt}\lambda_{Ru}$$

where t is the diagnosis time with intensity, λ_{Dt}, and u the reporting delay, with intensity, λ_{Ru}. This is just the independence log linear model, which can be fitted as

$$\texttt{DELAY} + \texttt{QUARTER} \tag{7.7}$$

where DELAY and QUARTER are appropriate factor variables. When there are missing cells, we have the quasi-independence model of Section 5.2. The triangle of missing data is weighted out in order to be able to obtain predictions of total incidence for the recent quarters, in the same way as we obtain population estimates in Section 5.4. A poor fit of this model would indicate non-stationarity of the delay intensities over time. For the data of Table 7.12, we obtain a deviance of 716.48 (820.48) with 413 d.f., indicating substantial non-stationarity (although parts of the table are rather sparse).

The fitted values for the diagnosis–time margin in this non-parametric stationary model are plotted as the broken line in Figure 7.5, showing the predicted incidence of AIDS rising sharply in 1992.

Because of the missing triangle, a complete non-stationary model, the saturated model of Equation (DELAY*QUARTER), cannot be fitted, and, in any case, is of little interest. One possible simple non-stationary model is the following interaction model:

$$\texttt{DELAY} + \texttt{QUARTER} + \texttt{DELAYL} \cdot \texttt{QUARTER} + \texttt{QUARTL} \cdot \texttt{DELAY} \tag{7.8}$$

where DELAYL and QUARTL are linear, instead of factor, variables. DELAYL uses centres of three month quarterly periods, but with 0.1 for the no reporting delay (to allow for logarithms below). Because we are interested in rates or intensities per unit time, an offset of log 0.2 months has (arbitrarily) been used for the zero reporting delay, as compared to log 3 months for all other delay periods, in all models. (These data are very robust to changes in the size of this first interval.) For our data, the deviance decreases by 131.09 (AIC 787.39) with 49 d.f. with respect to the stationary model, an even stronger indication of non-stationarity, because the sparseness should no longer be a major problem for a model so far from the saturated one. For future comparison, the deviances for these two non-parametric models are given in Table 7.13.

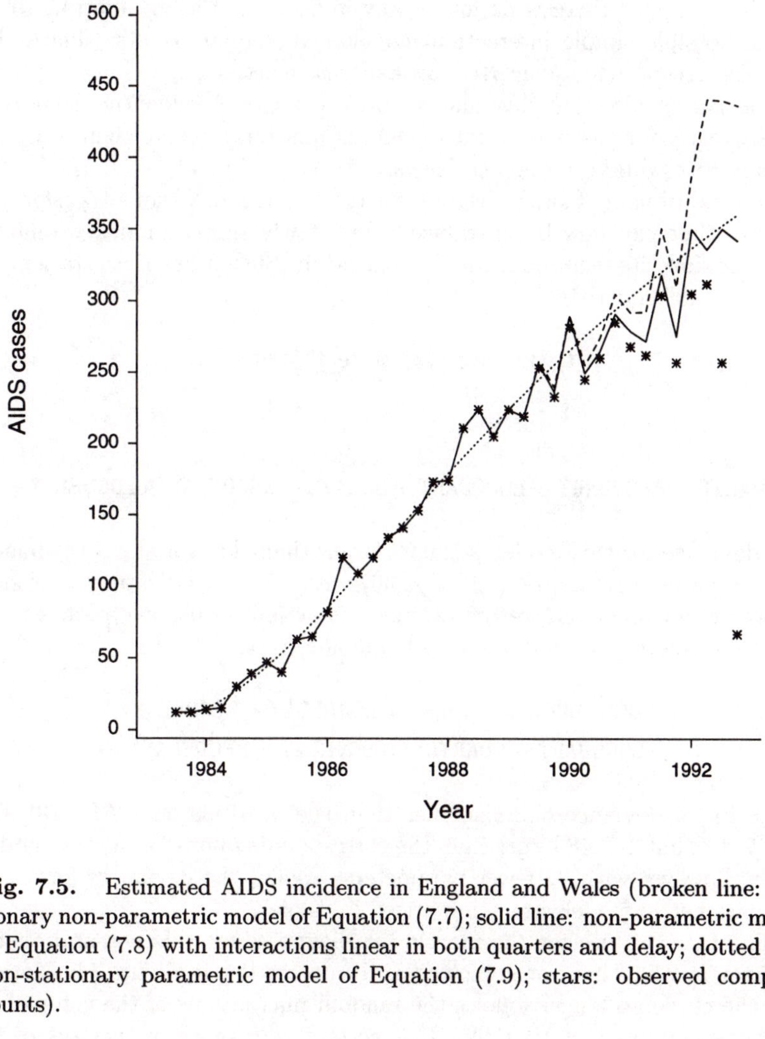

Fig. 7.5. Estimated AIDS incidence in England and Wales (broken line: stationary non-parametric model of Equation (7.7); solid line: non-parametric model of Equation (7.8) with interactions linear in both quarters and delay; dotted line: non-stationary parametric model of Equation (7.9); stars: observed complete counts).

Table 7.13. Deviances for a series of models fitted to the reporting-delay data of Table 7.12.

	Stationary		Non-stationary	
Model	Deviance	d.f.	Deviance	d.f.
Non-parametric	716.48 (820.48)	413	585.39 (787.39)	364
Parametric	837.40 (851.40)	458	773.87 (791.87)	456

This non-stationary model has been plotted as the solid line in Figure 7.5. We see that there is no longer any indication of a big jump in 1992. Other possible simple interaction models give similar results. Thus, the apparent recent increase in AIDS incidence is an artifact of an inappropriate model. Notice also that this, and the previous, model follow the diagnosed cases exactly for the period until 1989 (24 quarters), where there were no missing cases due to reporting delays.

Instead of using factor variables for the two times, various transformations of time can now be introduced. In a fairly simple model, we might consider time, its reciprocal, and its logarithm. Such a bivariate stationary model would be written

$$\lambda(t,u) = \alpha e^{\beta_1 t+\beta_2/t} t^{\beta_3} e^{\beta_4 u+\beta_5/u} u^{\beta_6}$$

or

$$\texttt{QUARTL} + \texttt{RECQUART} + \texttt{LOGQUART} + \texttt{DELAYL} + \texttt{RECDELAY} + \texttt{LOGDELAY}$$

The deviance for this model is 120.92 larger than that for the corresponding non-parametric model (AIC 851.40) with 45 d.f., indicating a considerably poorer model. If we try various interaction terms, to obtain a non-stationary model, we find one possible model to be

$$\begin{aligned} &\texttt{LOGQUART} + \texttt{RECQUART} + \texttt{LOGDELAY} + \texttt{RECDELAY} \\ &+(\texttt{LOGQUART} + \texttt{QUARTL}) \cdot (\texttt{LOGDELAY} + \texttt{RECDELAY}) \end{aligned} \tag{7.9}$$

which has a deviance 63.53 smaller than the previous one (AIC 791.87) with 2 d.f., but 188.48 larger than the corresponding non-parametric model above, however with 92 fewer parameters. Again, the deviances for these two parametric models are summarized in Table 7.13.

The fitted values for the margin for diagnoses for this non-stationary parametric model have been plotted as the dotted line in Figure 7.5. Notice how the curve no longer follows the random fluctuations of the completed diagnosis counts up until 1989. This accounts for much of the lack of fit of this model. The predictions, where there are missing reports, are very similar to those for the non-stationary, non-parametric model above, except for the smoothing.

7.10 Exercises

(1) Exercise (1.4) gave the numbers of suicides in the U.S.A. for 1968, 1969, and 1970.

   (a) Find an appropriate model using the techniques of this chapter, as well, perhaps, as those already used.

(b) Does the trend found in Section 1.4 hold for all three years analysed simultaneously?

(2) The age distribution of poliomyelitis cases was recorded for a small Inuit population during an epidemic in 1949 (Aickin, 1983, p. 105):

Age	0–4	5–9	10–14	15–19	20–29	30–49	50+
Population	53	56	33	26	30	52	25
Polio cases	2	13	8	11	6	12	5

Is the rate of cases occurring the same in all age groups?

(3) Cervical cancer deaths and thousands of woman–years at risk were recorded in four European countries between 1969 and 1973 (Whittemore and Gong, 1991):

	Age							
	25–34		35–44		45–54		55–64	
England and Wales	192	15399	860	14268	2762	15450	3035	15142
Belgium	8	2328	81	2557	242	2268	268	2253
France	96	15324	477	16186	998	14432	1117	13201
Italy	45	19115	255	18811	621	16234	839	15246

(a) Are there differences in mortality rate among countries?
(b) Is there evidence of evolution of the rate with age?
(c) If so, is it the same for all countries?

(4) The table below shows the monthly numbers of cases of poliomyelitis over 14 years in the U.S.A. (Zeger, 1988).

Year	J	F	M	A	M	J	J	A	S	O	N	D
1970	0	1	0	0	1	3	0	2	3	5	3	5
1971	2	2	0	1	0	1	3	3	2	1	1	5
1972	0	3	1	0	1	4	0	0	1	6	14	1
1973	1	0	0	1	1	1	1	0	1	0	1	0
1974	1	0	1	0	1	0	1	0	1	0	0	2
1975	0	1	0	1	0	0	1	2	0	0	1	2
1976	0	3	1	1	0	2	0	4	0	2	1	1
1977	1	1	0	1	1	0	2	1	3	1	2	4
1978	0	0	0	1	0	1	0	2	2	4	2	3
1979	3	0	0	2	7	8	2	4	1	1	2	4
1980	0	1	1	1	3	0	0	0	0	1	0	1
1981	1	0	0	0	0	0	1	2	0	2	0	0
1982	0	1	0	1	0	1	0	2	0	0	1	2
1983	0	1	0	0	0	1	2	1	0	1	3	6

(a) Is a homogeneous Poisson process suitable for these data?
(b) Is there any indication that the intensity of cases varies with month?

(c) Has the intensity changed over the years?

(5) The table below gives the number of occurrences (out of two) of rainfall over 1mm in Tokyo for the two years, 1983-84 (Kitagawa, 1987).

Jan.	00110	11000	00001	00110	11000	00000	1
Feb.	01000	00000	01100	02100	00110	1000	
Mar.	01100	00002	00110	21011	10120	01111	1
Apr.	20011	01201	11001	21021	01000	00100	
May.	11000	11000	00111	22000	00111	00010	0
Jun.	00010	10002	11210	12202	21112	22110	
Jul.	00101	11210	11002	11112	20100	00110	0
Aug.	00000	00000	00111	11101	21000	00001	0
Sep.	10001	01112	00012	20112	21011	11110	
Oct.	00001	00010	21101	11021	11100	01000	0
Nov.	00000	11011	90001	10011	00010	00000	
Dec.	00000	00000	10000	11100	00100	00001	1

(a) What would be a suitable model for these data?

(b) Is there any evidence of seasonal variation?

(6) The table below gives a series indicating if patients were arriving (1) at an intensive care unit each day from February 1963 to March 1964 (Lindsey, 1992, p. 26, from Cox and Lewis, 1966, pp. 254–255; read across rows).

00010	00100	10000	10101	10001	00110	10001	01000
00111	00101	01000	10100	10001	00111	00011	00000
01000	01100	00101	10001	01101	01110	11110	01010
10101	00001	01100	10100	11011	11011	01000	00111
01100	00001	10110	01010	01110	00100	01010	00001
01001	00000	01010	01011	01101	01101	00101	10011
00111	00101	00011	00000	11011	00100	01110	01111
11011	00111	11001	11011	01111	10101	11011	11111
00111	11100	10010	11011	10011	10110	10111	00110
00111	00001	11000	11000	01111	00111	10001	01010
00110	00000	1					

(a) Determine if the probability of patients arriving depends on previous arrivals.

(b) Is there a seasonal trend in the probability of patients arriving?

(7) (a) Use the techniques of this chapter to fit a counting process to the data of Exercise (6.13), taking into account the censoring.

(b) How do the results differ from those obtained in Chapter 6?

(8) The table below gives the number of operating hours between successive failures of air-conditioning equipment in ten aircraft (Cox and Lewis, 1966, p. 6, from Proschan).

Aircraft Number									
2	3	4	5	6	7	8	9	12	13
413	90	74	55	23	97	50	359	487	102
14	10	57	320	261	51	44	9	18	209
58	60	48	65	87	11	102	12	100	14
37	186	29	104	7	4	72	270	7	57
100	61	502	220	120	141	22	603	98	54
65	49	12	239	14	18	39	3	5	32
9	14	70	47	62	142	3	104	85	67
169	24	21	246	47	68	15	2	91	59
447	56	29	176	225	77	197	438	43	134
184	20	386	182	71	80	188		230	152
36	79	59	33	246	1	79		3	27
201	84	27	15	21	16	88		130	14
118	44	153	104	42	106	46			230
34	59	26	35	20	206	5			66
31	29	326		5	82	5			61
18	118			12	54	36			34
18	25			120	31	22			
67	156			11	216	139			
57	310			3	46	210			
62	76			14	111	97			
7	26			71	39	30			
22	44			11	63	23			
34	23			14	18	13			
	62			11	191	14			
	130			16	18				
	208			90	163				
	70			1	24				
	101			16					
	208			52					
				95					

(a) Compare possible event history models for these data.

(b) Are there differences among the risks of failure for the aircraft?

(9) The following table lists the days with accidents (1) for one shift in a mine (Lindsey, 1992, p. 21, from Barnett and Lewis, 1984, pp. 195; read across rows).

10010	00000	00000	00000	00000	01010	01011	11100
10110	10100	10010	00101	00100	00101	00010	00010
00101	11110	00000	01010	10101	01000	00001	00000
00101	11000	00001	10000	00010	01011	10010	10001
00100	00000	00001	00000	00001	00010	01010	10000
00000	00001	10001	00000	001			

(a) Determine if the probability of an accident depends on what happened on previous days.

(b) Is there any systematic trend in the probability of an accident?

(10) The following table gives the times (months) between recurrences of tumours for patients with bladder cancer under three treatments (Andrews and Herzberg, 1985, pp. 254–259).

Patient No.	Placebo								
6	6								
9	5								
10	12	4							
12	10	5							
13	3	13	7						
14	3	6	12						
15	7	3	6	8					
16	3	12	10						
18	1								
19	2	24							
20	25								
24	28	2							
25	2	15	5						
26	3	3	2	4	14				
27	12	3	9						
31	29								
33	9	8	5	2					
34	16	3	4	6	5	6			
36	3								
37	6								
38	3	3	3						
39	9	2	9	6	4				
40	18								
42	35								
43	17								
44	3	12	31	5	2				
46	2	13	9	6	4	5	4	6	3
47	5	9	5	8	14				
48	2	6	4	1	4	4	12	16	

Patient No.	Thiotepa								
83	5								
87	3								
88	1	2	2	2	3				
90	17								
91	2								
92	17	2							
97	6	6	1						
98	6								
99	2								
100	26	9							
102	22	1	4	5					
103	4	12	7	4	6	3	1		
104	24	2	3	11					
107	1	26							
109	2	18	3	4	11				
111	2								
115	4	20	23						
117	38								
Patient No.	Pyridoxine								
51	3	1							
53	2	1							
56	4								
57	3								
61	5								
64	3	7	12	4	8				
65	3	6	6	4	6				
67	3	4	5	4	3	9	6	2	3
70	2	4	4	6	7	4	9	3	3
72	10								
73	6	14							
74	8	7	3	2	2	3	13	2	
75	42								
77	44	3							
78	8	6	6	5	4	4	15	1	

(a) Does the risk of tumour recurrence depend on the number of previous tumours?

(b) Are there any differences in the risk of tumours under the three treatments?

(11) Laboratory animals were permitted 20 attempts to complete a task on each trial, stopping when 20 successes were reached (Aickin, 1983, p. 239). Four animals were placed in each of two combinations of stimulus (light or bell) and training.

Trial	Not trained							
	Light					Bell		
1	6	1	2	1	1	0	2	1
2	8	6	7	6	0	0	2	1
3	3	16	5	15	2	0	9	0
4	6	17	13	19	3	0	4	0
5	6	17	19	17	16	0	1	4
6	5	8	19	19	12	0	9	4
7	18	18	18	17	17	0	15	5
8	18	17	19	19	18	0	16	7
9	17	18	17	19	15	14	17	13
10	19	18	20	20	16	15	17	17
11	19	19	—	—	17	18	19	15
12	18	20	—	—	20	16	19	15
13	20	—	—	—	—	18	20	19
14	—	—	—	—	—	18	—	18
15	—	—	—	—	—	17	—	20
16	—	—	—	—	—	17	—	—
17	—	—	—	—	—	19	—	—
18	—	—	—	—	—	19	—	—
19	—	—	—	—	—	19	—	—
20	—	—	—	—	—	20	—	—

Trial	Trained							
	Light					Bell		
1	2	0	0	3	0	0	2	0
2	0	0	0	4	0	0	0	3
3	0	10	0	4	4	7	3	2
4	7	17	10	6	4	5	0	0
5	18	17	15	8	15	3	0	0
6	19	19	18	15	10	4	7	0
7	19	19	19	17	11	6	14	3
8	18	19	15	15	13	2	11	2
9	18	19	19	14	19	11	15	8
10	20	19	19	16	20	12	18	12
11	—	20	18	19	—	6	18	15
12	—	—	19	18	—	13	17	18
13	—	—	20	19	—	14	20	17
14	—	—	—	20	—	19	—	18
15	—	—	—	—	—	18	—	18
16	—	—	—	—	—	17	—	20
17	—	—	—	—	—	18	—	—
18	—	—	—	—	—	20	—	—

Fit an appropriate learning model to these data.

(12) The following table shows floating gallstone progression through time (days) to experience of biliary pain (first column) and to cholecystectomy (second column) under treatment and placebo (Wei and Lachin, 1984).

Placebo				Treatment			
741*	741*	35	118	735*	735*	742*	742*
234	234	175	493	29	29	360*	360*
374	733*	481	733*	748*	748*	750	750*
184	491	738*	738*	671	671	360*	360*
735*	735*	744*	744*	147	147	360*	360*
740*	740*	380	761*	749	749	726*	726*
183	740*	106	735*	310*	310*	727*	727*
721*	721*	107	107	735*	735*	725*	725*
69	743*	49	49	757*	757*	725*	725*
61	62	727	727*	63	260	288	810*
742*	742*	733*	733*	101	744*	728*	728*
742*	742*	237	237	612	763*	730*	730*
700*	700*	237	730*	272	726*	360*	360*
27	59	363	727*	714*	714*	758*	758*
34	729*	35	733*	282	734*	600*	600*
28	497			615	615*	743*	743*
43	93			35	749*	743*	743*
92	357			728*	728*	733*	755*
98	742*			600*	600*	188	762*
163	163			612	730*	600*	600*
609	713*			735*	735*	613*	613*
736*	736*			32	32	341	341
736*	736*			600*	600*	96	770*
817*	817*			750*	750*	360*	360*
178	727			617	793*	743*	743*
806*	806*			829*	829*	721*	721*
790*	790*			360*	360*	726*	726*
280	737*			96	720*	363	582
728*	728*			355	355	324	324
908*	908*			733*	733*	518	518
728*	728*			189	360*	628	628
730*	730*			735*	735*	717*	717*
721*	721*			360*	360*		

Asterisks indicate censoring.

(a) Can the same intensity function be used for time to biliary pain and to cholecystectomy?

(b) Is the transition rate between the two the same for both treatments?

(c) Does the risk of cholecystectomy depend on the time to biliary pain?

(13) The following table gives the times (sec.) at initiations of mating between flies (Aalen, 1978).

Ebony flies							
143	180	184	303	380	431	455	475
500	514	521	552	558	606	650	667
683	782	799	849	901	995	1131	1216
1591	1702	2212					
Oregon flies							
555	742	746	795	934	967	982	1043
1055	1067	1081	1296	1353	1361	1462	1731
1985	2051	2292	2335	2514	2570	2970	

(a) Look for an event history model to describe these data.

(b) Is there any difference between the two series for the two species of fly?

(14) In a capture–recapture study of Dunnocks (*Prunella modularis*), birds were captured, ringed, and released five times over a seven-month period (Cheke, 1985). Because of the ringing, it was known which birds were captured more than once, as shown in the following table:

Capture	Frequency	Capture	Frequency
YYYYY	0	NYYYY	1
YNYYY	0	NNYYY	0
YYNYY	0	NYNYY	1
YNNYY	0	NNNYY	1
YYYNY	0	NYYNY	0
YNYNY	0	NNYNY	1
YYNNY	0	NYNNY	0
YNNNY	0	NNNNY	14
YYYYN	1	NYYYN	1
YNYYN	0	NNYYN	2
YYNYN	1	NYNYN	2
YNNYN	1	NNNYN	16
YYYNN	0	NYYNN	2
YNYNN	0	NNYNN	11
YYNNN	2	NYNNN	13
YNNNN	10	NNNNN	—

(a) Estimate the total size of the population.

(b) What assumptions are being made about the birds being captured each time?

(c) How does the fact that the same birds are involved several times influence the results?

(15) The table below gives the numbers of AIDS cases reported to the Centers for Disease Control in the U.S.A. (Hay and Wolak, 1994), with reporting delays.

	Delay period (quarters)								
Year	0	1	2	3	4	5	6	7	8
1982	31	49	32	10	5	10	5	2	4
	40	67	11	5	10	9	7	3	0
	78	73	32	21	12	11	1	3	1
	96	129	30	33	17	5	2	3	1
1983	134	177	68	34	14	12	4	7	3
	57	378	85	43	20	18	12	9	5
	69	420	113	34	19	12	10	10	4
	26	513	109	55	25	17	7	8	4
1984	55	675	151	59	32	26	18	8	9
	82	790	164	85	57	36	16	4	11
	108	845	241	112	47	40	18	16	15
	118	960	247	112	65	30	27	15	11
1985	146	1191	252	129	83	67	34	20	18
	160	1454	292	143	93	58	48	35	24
	152	1620	400	225	101	71	53	39	20
	97	1739	422	164	120	58	52	52	57
1986	148	2046	406	218	107	118	56	107	135
	362	2039	555	200	143	91	152	160	133
	232	2444	532	275	148	196	229	165	123
	181	2441	763	290	240	282	183	143	101
1987	224	2981	673	408	370	353	224	185	99
	129	3260	897	592	426	272	193	125	133
	96	3567	1207	569	374	227	156	138	102
	135	3847	1218	444	315	247	195	128	149
1988	163	4401	1096	462	354	334	196	186	203
	307	4608	968	500	372	284	225	231	182
	332	4521	1186	569	334	317	222	193	—
	256	4525	1327	487	375	387	268	—	—
1989	311	5016	1248	569	527	438	—	—	—
	342	5186	1370	814	512	—	—	—	—
	349	5124	1515	830	—	—	—	—	—
	192	4998	1745	—	—	—	—	—	—
1990	276	5646	—	—	—	—	—	—	—
	329	—	—	—	—	—	—	—	—

	Delay period (quarters)							
Year	9	10	11	12	13	14	15	16+
1982	0	2	0	0	0	0	0	35
	2	1	3	1	0	1	0	41
	2	2	1	1	1	2	2	50
	0	2	3	1	0	0	1	58
1983	4	3	4	2	1	2	0	67
	6	5	0	5	5	2	3	52
	4	3	4	3	3	7	4	50
	3	7	9	8	7	5	0	48
1984	7	7	4	9	7	5	6	70
	9	6	12	9	11	5	11	65
	9	8	8	5	13	11	7	70
	18	15	13	13	16	11	15	60
1985	22	10	18	22	27	29	21	68
	20	29	46	33	31	23	27	62
	56	55	44	29	35	22	21	54
	65	83	41	37	27	29	17	47
1986	102	81	49	53	40	26	30	53
	94	77	66	41	31	27	38	54
	80	62	58	36	42	39	31	—
	82	67	35	38	50	39	—	—
1987	126	85	85	78	56	—	—	—
	121	91	96	74	—	—	—	—
	118	117	85	—	—	—	—	—
	140	102	—	—	—	—	—	—
1988	165	—	—	—	—	—	—	—
	—	—	—	—	—	—	—	—
	—	—	—	—	—	—	—	—
	—	—	—	—	—	—	—	—
1989	—	—	—	—	—	—	—	—
	—	—	—	—	—	—	—	—
	—	—	—	—	—	—	—	—
	—	—	—	—	—	—	—	—
1990	—	—	—	—	—	—	—	—
	—	—	—	—	—	—	—	—

(a) Fit a bivariate intensity function to these data.

(b) Plot the estimated AIDS incidence from your models.

(c) How do predictions from the stationary model differ from those in Figure 7.5?

(d) Plot three-dimensional graphs of your best parametric model and of that for the data in Table 7.12.

(e) Contrast the evolution of reporting delays in the two countries.

8
Markov chains

In Section 7.3, we briefly introduced models for Markov chains in the simple case where there were only two possible responses: an event occurs or not. However, such models have much wider application. In continuous-time models, where each subject is in one of several possible states at any given point, they are often called semi-Markov processes or Markov renewal processes and may be analysed by the counting process methods of Chapter 7. The intensities, then, refer to the transition probabilities between states, as we saw in Section 7.7.

In this chapter and the next, we shall be looking at Markov chains in discrete time, where the response is only recorded a few times, but where several states are usually possible. In such cases, observations are usually made on a large number of units. Here, we shall consider some traditional aspects of such chains, such as their order and whether or not the transition matrix is stationary. In the next chapter, we study how to impose more complex structures on the transition matrix.

8.1 Retrospective studies

In a social mobility study, as in many related social studies, we obtain a sample of people with their characteristics, and then retrospectively obtain information about their parents. For social mobility, the information is specifically about occupation, but the same principle applies for education, political beliefs, and so on, as we saw in Chapter 2.

As the name implies, a retrospective mobility study does things backwards. We have a certain number of children (almost invariably sons) of each occupational category, and we look back to see from which parental (father's) occupational category they came. We have a sampling structure which implies that we can calculate the probability of the father having any given occupation given the son's occupation. This is the exact opposite of what we want. In addition, our sample, if correctly chosen, will be representative of the sons' occupations but not of the fathers'. The occupational and demographic structure may have changed between the two generations, but we are interested in mobility, not in these structural changes.

Our observations take the form of a two-way table, cross-classifying the two occupational variables. Our two problems, the retrospective nature of

Table 8.1. British inter-generational social mobility. The categories are A. Professional, high administrative, B. Managerial, executive, high supervisory, C. Low inspectional, supervisory, D. Routine non-manual, skilled manual, and E. Semi- and unskilled manual. (Bishop *et al.*, 1975, p. 100, from Glass)

	Father				
Son	A	B	C	D	E
A	50	45	8	18	8
B	28	174	84	154	55
C	11	78	110	223	96
D	14	150	185	714	447
E	3	42	72	320	411

the study and the changes in occupational structure, may be resolved by the same procedure: we study changes within the table, given (conditional on the fact) that both marginal totals are fixed; we then apply the log linear model. It is possible to demonstrate that this is the only procedure which can resolve these two problems. On the other hand, the transition matrix cannot be correctly estimated in such a retrospective study, for the reasons given above.

We shall now briefly consider a classical social mobility table, as shown in Table 8.1. More detailed analyses of similar mobility tables will be given in Chapter 9, where a number of more complex models for such studies will be introduced. Here, we shall only consider whether or not the son's occupation depends on the father's. If it does not, the two variables are independent.

We first fit the model where only the two sets of marginal frequencies are fixed:

$$\log(\nu_{ij}) = \mu + \theta_i + \phi_j \tag{8.1}$$

where i indexes the son's occupation and j the father's, or correspondingly

```
SON + FATHER
```

This is the model for independence between the two occupational situations. The results show that this model cannot be accepted: the deviance is 792.19 (810.19) with 16 d.f. A glance at the residuals (not shown) reveals that all but one of the diagonal values are large and positive; the model has underestimated these values. This model of independence does not predict the observed fact that many sons remain in the same occupational category as their fathers. This is reflected in Cook's distances, in Figure 8.1, for observations 1, 7, 19, and 25. The normal probability plot, in Figure

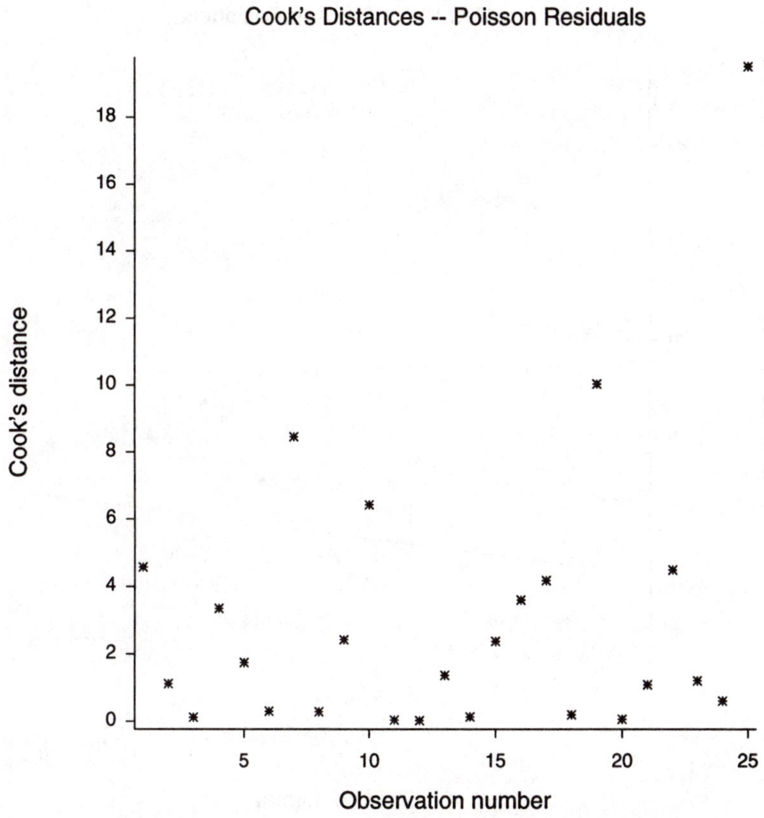

Fig. 8.1. Cook's distances from the independence model of Equation (8.1) for the social mobility data of Table 8.1.

8.2, indicates a poor fit.

As we saw in Chapter 2, an alternative is to fit the saturated model, a model with as many parameters as there are entries in the table:

$$\log(\nu_{ij}) = \mu + \theta_i + \phi_j + \gamma_{ij}$$

which corresponds to

`FATHER * SON`

This model must necessarily fit the data exactly. As we know, γ_{ij} represents a matrix of parameters describing exactly the observed mobility between generations under the conditions set forth above: a retrospective study with changing occupational structure.

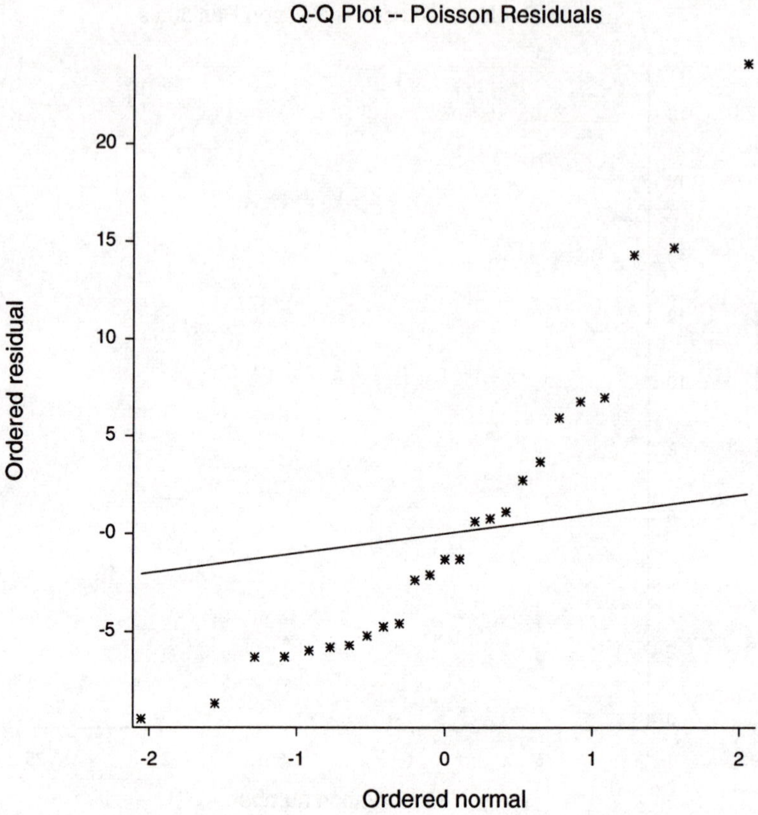

Fig. 8.2. The normal probability plot from the independence model of Equation (8.1) for the social mobility data of Table 8.1.

As would be expected, the deviance is zero (50). The matrix of parameter values, γ_{ij}, using the conventional constraints, is

$$\begin{array}{rrrrr} 2.467 & 0.447 & -0.379 & -0.961 & -1.574 \\ 0.623 & 0.536 & -0.159 & -0.328 & -0.672 \\ -0.832 & 0.080 & 0.458 & 0.154 & 0.139 \\ -1.014 & -0.307 & 0.171 & 0.512 & 0.638 \\ -1.245 & -0.756 & -0.091 & 0.624 & 1.469 \end{array}$$

We see from this matrix once again how the diagonal categories are overrepresented. Members of the two extreme categories, professional and high administrative and semi- and unskilled have especially little mobility. We now have values measuring dependence between fathers' and sons' occupations which eliminate the bias from the retrospective method and from

Table 8.2. Members of the leading crowd at two points in time. (Coleman, 1964, p. 171)

			Attitude 2	
Attitude 1	Member 1	Member 2	Favourable	Unfavourable
Favourable	Yes	Yes	458	140
Unfavourable			171	182
Favourable	No		184	75
Unfavourable			85	97
Favourable	Yes	No	110	49
Unfavourable			56	87
Favourable	No		531	281
Unfavourable			338	554

changes in occupational structure.

In Chapter 9, we shall study other intermediate, more structured, models between the independence and saturated models for such square tables as this mobility table.

8.2 Prospective studies

As we saw in Section 2.3, a prospective study follows subjects over time. If the same set of questions is posed to the same individuals at several points in time, it is often called a panel study. Here, we necessarily avoid the time-sequential ambiguity of the example in Section 2.4, at least for relations between time points. The ambiguity may still remain for relationships among different responses at the same time point.

In this section, we shall consider a classical example of a two-wave panel with two variables at each time point. This is a study of the social system of schoolboys in ten high schools in the U.S.A. It looked in particular at membership in the 'leading crowd', and attitudes to it. Questions included one's self-perceived membership in it and whether being a member would imply sometimes going against one's principles. The questionnaire was administered in October 1957, and May 1958, with results given in Table 8.2.

In contrast to the previous example, all variables are dichotomous, so that we might use the logistic model with the binomial distribution. However, there is no obvious order between the variables in a wave, so that it is more reasonable to consider a bivariate distribution at each time point, using log linear models, as we did in Section 2.6 for one point in time. This will involve examining the table twice, the first time after excluding the second wave variables by collapsing it over them.

In the example of Section 2.4, we studied all three variables simultaneously in the same model, because any one could possibly influence the others. Here, the responses of the second time cannot influence those of the first so that we must analyse them in a model which does not include the second wave.

In general, in a panel study, it will not be possible to distinguish direction of causality within a wave. In our case, self-evaluation of membership in the leading crowd and attitude to that crowd may be mutually influential. Thus, in the first wave, we take attitude towards the leading crowd and membership as mutually interacting in the log linear model,

$$\texttt{MEM1} * \texttt{ATT1}$$

The relationship is strong, with a parameter estimate of 0.106, an AIC of 8, and a deviance of 35.16 (41.16) with 1 d.f. for the corresponding independence model. Members have a more favourable attitude than do non-members, believing that membership does not involve going against principles.

We now take, in our second model, the second membership and attitude as responses. First, we fit the null model with no dependence of the two responses on the explanatory variables

$$\texttt{MEM2} + \texttt{ATT2} + \texttt{MEM1} * \texttt{ATT1} \tag{8.2}$$

The deviance of 1386.5 (1398.5) with 10 d.f. indicates that membership and attitude at the second point in time depend on at least some of the other variables and/or on each other. Cook's distances in Figure 8.3 show that the first and last categories are least well fitted by the model; the members with favourable attitude at both time points, and non-members who are unfavourable both times, are underestimated.

Next, we fit a model with the main effects, attitude and membership at point one, on the two responses, as well as an association between the two responses:

$$\texttt{MEM2} * \texttt{ATT2} + \texttt{MEM1} * \texttt{ATT1} + (\texttt{MEM1} + \texttt{ATT1}) * (\texttt{MEM2} + \texttt{ATT2}) \tag{8.3}$$

This model fits well, with a deviance of 1.21 (23.21) and 5 d.f.; no interaction effects are necessary between the two explanatory variables with respect to the two responses. Cook's distances in Figure 8.4 indicate that the problem with the extremes has virtually disappeared.

We may now wonder if any of the main effects might be eliminated. Instead of removing each variable in turn, let us look at the relationships between the parameter estimates and their standard errors, as given in Table 8.3. We see that the ratio for attitude at point two depending on membership at point one is considerably less than 2: (0.038/0.0225), which is not the case for the other relationships. We shall try to eliminate it. The

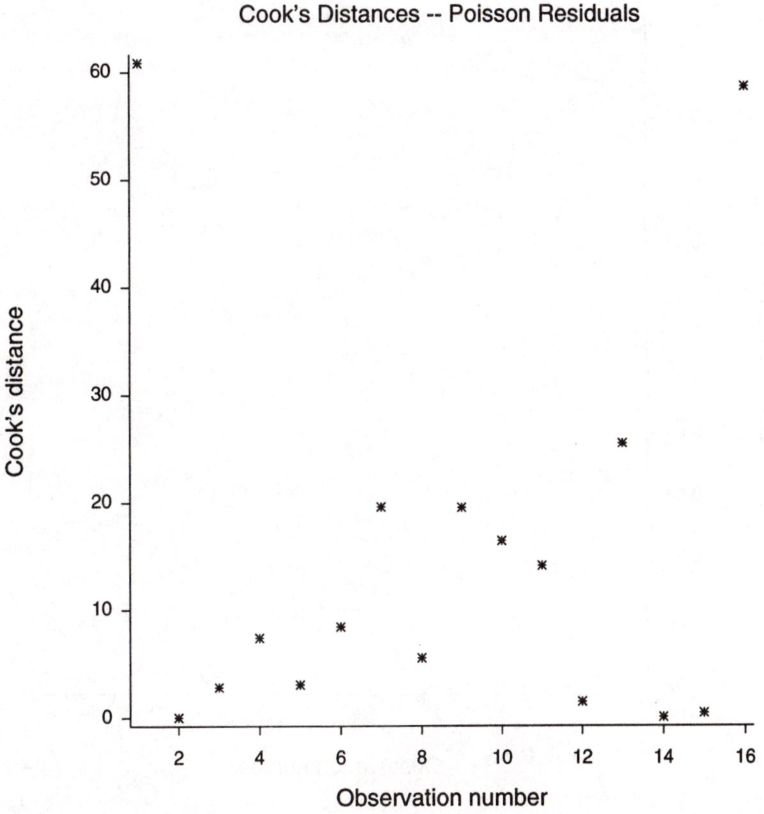

Fig. 8.3. Cook's distances from the minimal model of Equation (8.2) at the second wave for the leading crowd data of Table 8.2.

Table 8.3. Parameter estimates, with attitude at time 2 as response, for the leading crowd data of Table 8.2.

	Estimate	s.e.
MEM1·MEM2	0.617	0.0214
ATT1·MEM2	0.086	0.0219
MEM1·ATT2	0.038	0.0225
ATT1·ATT2	0.290	0.0182
MEM2·ATT2	0.084	0.0221

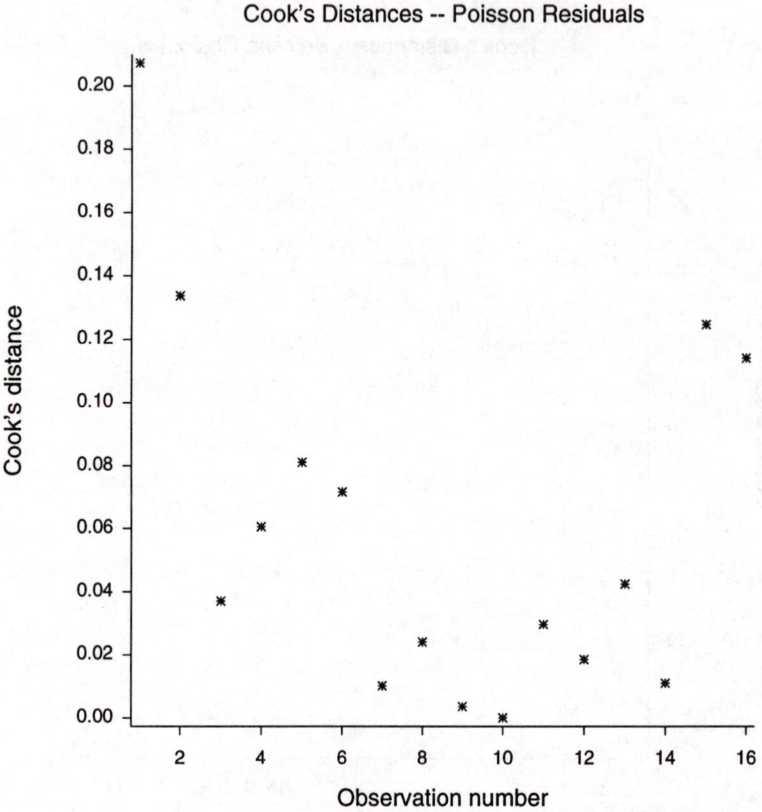

Fig. 8.4. Cook's distances from the complete main effects model of Equation (8.3) at the second wave, for the leading crowd data of Table 8.2.

deviance, 4.06 (24.06), is now larger, but not too much; either model may be acceptable. Neither of the other variables can be eliminated.

Membership at the second wave depends very strongly on previous membership and is slightly more probable when previous attitude was favourable. Attitude is more favourable at point 2 if one was a member and if attitude was favourable previously. As might be expected, previous membership has little effect on current attitude when these other variables, including present membership, are taken into account. Finally, attitude and membership at time 2 are positively associated.

We may now summarize our findings in a path diagram, or interaction graph, as in the example of Section 2.4:

	Time	
1		2
Membership	→	Membership
↕	↗	↕
Attitude	→	Attitude

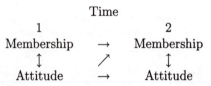

In general, current membership and current attitude are positively related. Previous membership and attitude affect current membership positively, while previous attitude (and previous membership very little) is related to current attitude. As before, such a diagram should not be interpreted as indicating causality.

8.3 Stationarity of Markov chains

When panel observations are available over more than two time periods, it is possible to determine if the same pattern of change occurs in each period. Suppose that individual responses at a given time point depend only on those of the immediately preceding point. As we have seen, this is the hypothesis of a first-order Markov chain. Then, the probability of an individual belonging to any given category depends only on his/her category for the immediately preceding time point. We have a square transition matrix of probabilities, as in the previous sections. If the rows are the categories at the previous time point and the columns are the present categories, then the row probabilities sum to one. This matrix represents the pattern of change; if it is the same over each period, we have a stationary transition matrix. In this section, we shall look at this stationarity of a first-order Markov chain, assuming the first-order hypothesis. In the next section we shall look at the latter hypothesis. Another aspect of stationarity is when the marginal distribution of states remains constant over time. We look at that in Section 9.2.

A common application of Markov chains is to voting behaviour. Here, we consider successive monthly expressions of intention to vote in the 1940 U.S.A. presidential elections for Erie County, as given in Table 8.4. The data consist of a series of five two-way tables, yielding a three-way table over the five time periods.

We have three variables: the voting intention at the beginning of any period (three categories — `BEGIN`), the voting intention at the end of any period (three categories — `END`), and the time periods themselves (five categories — `PERIOD`). The model for stationarity is one with independence between time period and intention to vote at the end of the period; this relationship is omitted from the model:

$$\mathtt{END} * \mathtt{BEGIN} + \mathtt{PERIOD} * \mathtt{BEGIN} \tag{8.4}$$

Thus, the model will contain the three sets of mean parameters and those for the relationships between intentions at the beginning and end of a period

Table 8.4. One-step transitions for voting intentions in the 1940 U.S.A. presidential elections, Erie County. (Goodman, 1962)

	Party	Republican	Democrat	Undecided
			June	
May	Republican	125	5	16
	Democrat	7	106	15
	Undecided	11	18	142
			July	
June	Republican	124	3	16
	Democrat	6	109	14
	Undecided	22	9	142
			August	
July	Republican	146	2	4
	Democrat	6	111	4
	Undecided	40	36	96
			September	
August	Republican	184	1	7
	Democrat	4	140	5
	Undecided	10	12	82
			October	
September	Republican	192	1	5
	Democrat	2	146	5
	Undecided	11	12	71

and for those between intentions at the beginning of a period and the period itself. The transition matrix (assuming stationarity) can be calculated as

$$\begin{matrix} 0.928 & 0.014 & 0.058 \\ 0.037 & 0.900 & 0.063 \\ 0.132 & 0.122 & 0.747 \end{matrix}$$

With a deviance of 101.51 (143.51) and 24 d.f., we see that the hypothesis of stationarity is decisively rejected. The process of changing intentions to vote varies over the months. Cook's distances in Figure 8.5 indicate that the diagonal elements for the first two periods and those for Democrat to Undecided and Republican to Democrat for the last two periods are poorly estimated. From the residuals, given in Table 8.5, we see that stability within the two parties is overestimated in the first two periods and underestimated in the last two. We may conclude that intentions are more stable in the last two periods because the diagonal (no change) is larger then, and that the main differences in changes in intentions between the first and last two periods are the two just mentioned.

Table 8.5. Fitted values and residuals from the stationarity model of Equation (8.4) for the one-step voting intention data of Table 8.4.

			Observed	Fitted	Residual
May–June	Rep	Rep	125	135.5	−0.899
		Dem	5	2.1	1.992
		Und	16	8.4	2.606
	Dem	Rep	7	4.7	1.058
		Dem	106	115.2	−0.857
		Und	15	8.1	2.427
	Und	Rep	11	22.5	−2.426
		Dem	18	20.8	−0.621
		Und	142	127.7	1.270
June–July	Rep	Rep	124	132.7	−0.753
		Dem	3	2.1	0.651
		Und	16	8.3	2.693
	Dem	Rep	6	4.7	0.577
		Dem	109	116.1	−0.659
		Und	14	8.2	2.046
	Und	Rep	22	22.8	−0.163
		Dem	9	21.1	−2.631
		Und	142	129.1	1.131
July–Aug	Rep	Rep	146	141.0	0.419
		Dem	2	2.2	−0.132
		Und	4	8.8	−1.613
	Dem	Rep	6	4.4	0.736
		Dem	111	108.9	0.201
		Und	4	7.7	−1.320
	Und	Rep	40	22.6	3.647
		Dem	36	21.0	3.286
		Und	96	128.4	−2.859
Aug–Sept	Rep	Rep	184	178.1	0.439
		Dem	1	2.8	−1.065
		Und	7	11.1	−1.228
	Dem	Rep	4	5.5	−0.631
		Dem	140	134.1	0.509
		Und	5	9.4	−1.441
	Und	Rep	10	13.7	−0.998
		Dem	12	12.7	−0.189
		Und	82	77.6	0.495
Sept–Oct	Rep	Rep	192	183.7	0.612
		Dem	1	2.9	−1.100
		Und	5	11.4	−1.903
	Dem	Rep	2	5.6	−1.528
		Dem	146	137.7	0.707
		Und	5	9.7	−1.503
	Und	Rep	11	12.4	−0.391
		Dem	12	11.5	0.161
		Und	71	70.2	0.099

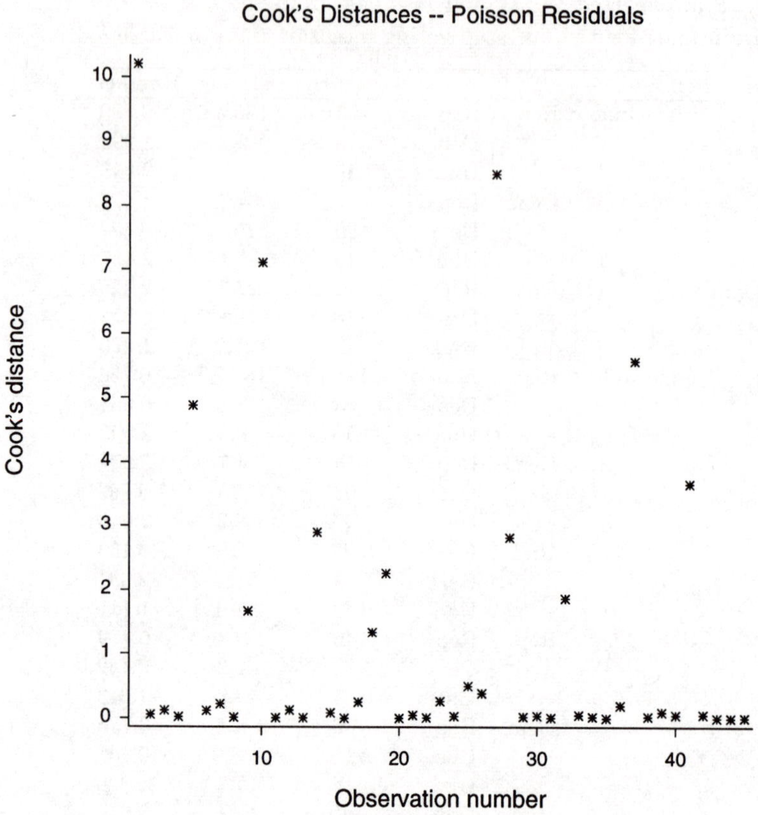

Fig. 8.5. Cook's distances from the stationarity model of Equation (8.4) for the one-step voting intention transition data of Table 8.4.

One important supplementary piece of information is that the Democratic convention was held during the third period. We may reconstruct separate tables for the first two time periods and for the last two and apply the model successively to each. With deviances of 7.41 (31.41) and 1.50 (25.50), each with 6 d.f., stationarity is no longer rejected in either separate table. The two transition matrices are now, respectively,

0.862	0.028	0.111
0.051	0.837	0.113
0.096	0.078	0.826

and

0.964	0.005	0.031
0.020	0.947	0.033
0.106	0.121	0.773

Table 8.6. Two-step transitions for voting intentions in the 1940 U.S.A. presidential elections, Erie County. (Goodman, 1962)

		Time $t+2$		
Time $t+1$	Time t	Republican	Democrat	Undecided
	Republican	557	6	16
Republican	Democrat	18	0	5
	Undecided	71	1	11
	Republican	3	8	0
Democrat	Democrat	9	435	22
	Undecided	6	63	6
	Republican	17	5	21
Undecided	Democrat	4	10	24
	Undecided	62	54	346

We note, as expected, that the diagonal transition probabilities are considerably smaller before the convention than after. In the first table (May–July), 86% of those intending to vote for the Republicans and 84% of those for the Democrats are estimated as not having changed their minds over a period of one month, whereas, after the convention (August–October), the percentages are 96 and 95 respectively. Cook's distances (not shown) no longer exhibit any consistent trend.

Remember, however, that this analysis supposes that intentions at one point in time only depend on intentions one month before. This is the hypothesis of a first-order Markov chain which will be examined in the next section.

8.4 Order of Markov chains

In order to look at whether a series of observations is a first-order Markov chain or one of higher order, we require the details of changes for units over successive periods, and not only over one period at a time, as in the preceding section. We can then simply check if the categories at the present time point are independent of those two time periods before. If so, the process is of first-order. If not, it is at least of second order.

We consider again the data from the same study as in the previous example. This time, we (wrongly) assume stationarity of second order and collapse the tables over the four three-month periods (May–July, June–August, July–September, August–October), as in Table 8.6. This table cannot be obtained from the one in the previous section which only covers two-month periods. Note that we could, and should, also look at second-order stationarity in a way similar to that of the preceding section. However, only the tables used in these two sections are provided in Goodman (1962), so that it is not possible here.

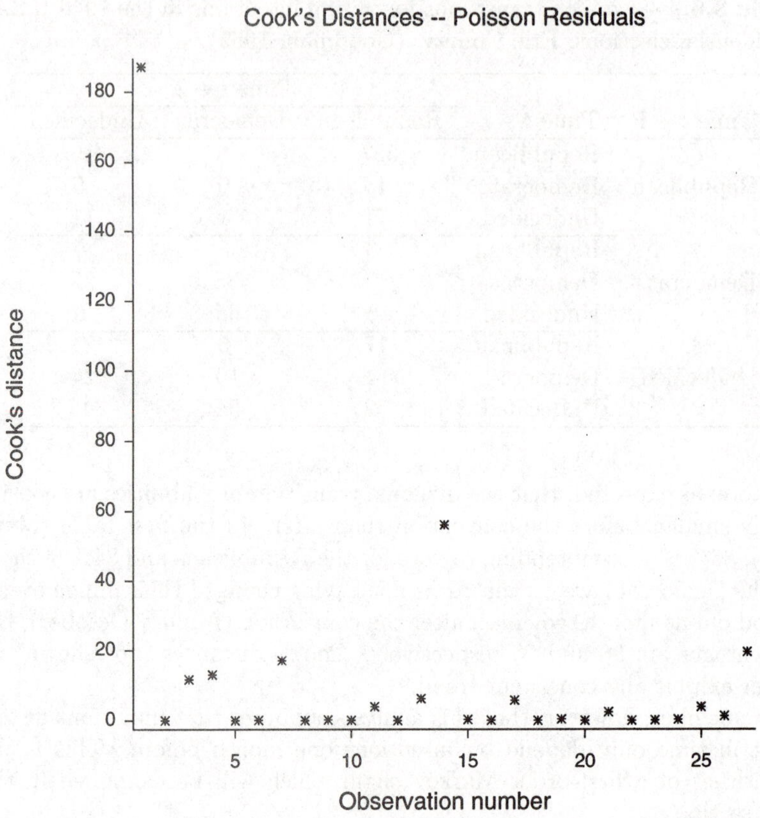

Fig. 8.6. Cook's distances from the first-order Markov chain model of Equation (8.5) for the two-step voting intention transition data of Table 8.6.

More concretely, we are assuming that the sequence of changes in voting intentions is identical over any consecutive three-month period (stationarity) and we check if current intentions in any month depend only on intentions in the previous month (first-order) or if they also depend on intentions two months before (second order).

We have a three-way table with intentions to vote at times t (T0), $t+1$ (T1), and $t+2$ (T2). We check to see if intentions at time $t+2$ are independent of those at time t:

$$\text{T2} * \text{T1} + \text{T1} * \text{T0} \tag{8.5}$$

With a deviance of 63.50 (93.50) and 12 d.f., the model is rejected. Cook's distances, in Figure 8.6, show the three cases of complete stability, espe-

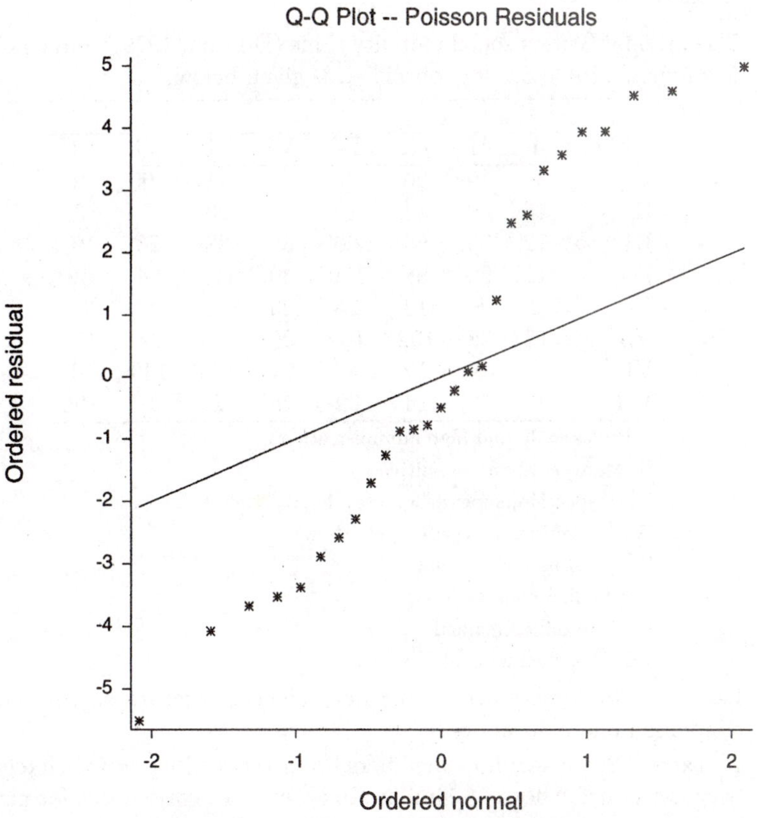

Fig. 8.7. The normal probability plot from the first-order Markov chain model of Equation (8.5) for the two-step voting intention transition data of Table 8.6.

cially the Rep·Rep·Rep cell, to be badly fitted by the model. As expected with a large deviance, the normal probability plot in Figure 8.7 deviates greatly from the 45 degree line. But, in addition, the residuals do not form a straight line, indicating that the second-order model assuming stationarity is poorly chosen. The deviation from 45 degrees results primarily from the poor fit of a second-order model, while non-linearity of the residual plot arises primarily from lack of stationarity.

Given stationarity, present voting intentions depend on more than just those of the preceding time point. They depend, at least, on the two previous points. With sufficient data, higher order Markov hypotheses could be tested in the same manner.

8.5 Exercises

(1) The original British social mobility table (Duncan, 1979, from Glass), from which Table 8.1 was obtained, is given below.

	Son							
Father	I	II	III	IV	Va	Vb	VI	VII
I	50	19	26	8	7	11	6	2
II	16	40	34	18	11	20	8	3
III	12	35	65	66	35	88	23	21
IV	11	20	58	110	40	183	64	32
Va	2	8	12	23	25	46	28	12
Vb	12	28	102	162	90	554	230	177
VI	0	6	19	40	21	158	143	71
VII	0	3	14	32	15	126	91	106

I: Professional and high administration
II: Managerial and executive
III: Inspection, supervisory, etc. (high)
IV: Inspection, supervisory, etc. (low)
Va: Routine non-manual
Vb: Skilled manual
VI: Semiskilled manual
VII: Unskilled manual

Does this table show the same dependence structure as the more condensed one used above?

(2) In Table 8.2, we saw how membership in the leading crowd changed over the year for boys. A similar study was also carried out for girls, as shown below (Coleman, 1964, p. 168):

			Attitude 2	
Attitude 1	Member 1	Member 2	Fav.	Unfav.
Favourable	Yes	Yes	484	93
Unfavourable			112	110
Favourable	No		129	40
Unfavourable			74	75
Favourable	Yes	No	107	32
Unfavourable			30	46
Favourable	No		768	321
Unfavourable			303	536

Are the same relationships found here as we saw for the boys above?

(3) Hourly transition frequencies were recorded in a study of the behaviour of male beavers in summer (Rugg and Buech, 1990):

Feed	Swim	Lodge	Other	Feed	Swim	Lodge	Other
	1800-1900				1900-2000		
2	0	0	0	176	15	5	0
2	13	0	0	25	281	2	2
0	0	0	0	0	0	24	0
0	0	0	0	2	3	0	7
	2000-2100				2100-2200		
503	32	1	10	321	34	0	1
35	243	1	7	34	292	3	3
0	6	149	0	0	2	76	0
6	10	1	35	1	3	1	24
	2200-2300				2300-0000		
156	12	0	2	66	11	0	0
14	82	0	1	9	81	1	2
0	0	9	0	0	1	6	0
2	1	0	5	0	0	0	50
	0000-0100				0100-0200		
41	2	0	2	45	8	0	1
5	26	1	3	5	50	1	3
0	1	3	0	0	0	6	0
1	3	0	38	1	1	0	4
	0200-0300				0300-0400		
16	3	0	0	28	2	0	0
1	39	0	0	2	21	0	0
0	1	1	0	0	0	0	0
0	0	0	0	0	0	0	0
	0400-0500				0500-0600		
0	0	0	0	11	0	0	0
0	0	0	0	0	15	0	0
0	0	0	0	0	1	24	0
0	0	0	0	0	0	0	0
	0600-0700				0700-0900		
78	3	5	2	33	1	3	1
9	63	6	4	1	19	2	5
0	12	107	2	0	4	40	2
1	5	3	58	0	2	5	24

(a) Are the transition matrices stationary over time?

(b) If not, is there any systematic change with different periods of the day?

(4) A sample of voting intentions was recorded from the same people in Holland in February and March, 1977, along with the age category of each person (Hagenaars, 1990, p. 178):

	Christian Democrat	Left	Other
February	March		
	Young		
Christian Democrat	90	10	10
Left	4	196	15
Other	20	24	220
	Old		
Christian Democrat	152	7	15
Left	5	154	11
Other	23	27	117

Does the way voting intentions change between months differ between the two age groups?

(5) The voting intention data in Holland between February and March 1977 (Exercise 8.4 above) were also classified by preference for Prime Minister (Hagenaars, 1990, p. 171). The possible Prime Ministers were Van Agt (Christian Democrat), Den Uyt (Left), and other (O).

		Preference for PM								
	Feb.	CD	CD	CD	L	L	L	O	O	O
	Mar.	CD	L	O	CD	L	O	CD	L	O
Vote Feb.	Mar.									
CD	CD	84	9	23	6	13	7	24	8	68
	Left	0	1	0	0	8	1	2	2	3
	O	3	1	2	0	2	3	2	3	9
Left	CD	1	1	0	1	2	2	1	0	1
	Left	2	4	0	1	293	6	1	22	21
	O	1	0	0	1	8	7	0	0	9
O	CD	6	1	1	4	5	0	9	1	16
	Left	0	1	1	0	31	0	2	9	7
	O	14	1	15	3	48	23	12	21	200

Study how the relationship between voting intention and preference for a Prime Minister changed over the two months.

(6) The following table shows the transitions for ten years of daily rainfall in January in Durban, South Africa (Pegram, 1980):

	Days 2, 3, 4							
Days 0, 1	000	001	010	011	100	101	110	111
00	18	17	9	14	10	5	9	13
01	10	5	1	5	11	3	10	8
10	16	8	7	8	5	2	5	4
11	14	10	6	4	11	7	7	8

What order of Markov chain is necessary to describe these data?

(7) Changes in votes were recorded for a sample of people in the 1964, 1968, and 1970 Swedish elections (Fingleton, 1984, p. 151):

		1970			
1964	1968	SD	C	P	Con
SD	SD	812	27	16	5
	C	5	20	6	0
	P	2	3	4	0
	Con	3	3	4	2
C	SD	21	6	1	0
	C	3	216	6	2
	P	0	3	7	0
	Con	0	9	0	4
P	SD	15	2	8	0
	C	1	37	8	0
	P	1	17	157	4
	Con	0	2	12	6
Con	SD	2	0	0	1
	C	0	13	1	4
	P	0	3	17	1
	Con	0	12	11	126

SD — Social Democrat C — Centre
P — People's Con — Conservative

(a) Construct an appropriate Markov chain model for these data.
(b) Check for order and stationarity.
(c) How can the model be improved to take into account the unequal periods between elections?

(8) In a hypothetical example, a study was made to compare responses to two drugs during four weeks for patients having two different diagnoses (Koch *et al.*, 1977):

	Week 1	N	N	N	N	A	A	A	A
	Week 2	N	N	A	A	N	N	A	A
	Week 4	N	A	N	A	N	A	N	A
Diagnosis	Treatment								
Mild	Standard	16	13	9	3	14	4	15	6
	New	31	0	6	0	22	2	9	0
Severe	Standard	2	2	8	9	9	15	27	28
	New	7	2	5	2	31	5	32	6

N: Normal, A: Abnormal.

Develop a suitable model for these data.

(9) Mothers of young children living within ten miles of the Three Mile Island nuclear power plant were interviewed four times after the nuclear accident in the spring of 1979, in winter 1979, spring 1980, au-

tumn 1980, and autumn 1982, to determine stress levels (Conaway, 1989, from Fienberg *et al.*). The stress levels were classed as low (L), medium (M), and high (H), and the mothers were classified by distance from the plant:

			< 5 miles			> 5 miles		
Wave			Wave 4					
1	2	3	L	M	H	L	M	H
L	L	L	2	0	0	1	2	0
		M	2	3	0	2	0	0
		H	0	0	0	0	0	0
	M	L	0	1	0	1	0	0
		M	2	4	0	0	3	0
		H	0	0	0	0	0	0
	H	L	0	0	0	0	0	0
		M	0	0	0	0	0	0
		H	0	0	0	0	0	0
M	L	L	5	1	0	4	4	0
		M	1	4	0	5	15	1
		H	0	0	0	0	0	0
	M	L	3	2	0	2	2	0
		M	2	38	4	6	53	6
		H	0	2	3	0	5	1
	H	L	0	0	0	0	0	0
		M	0	2	0	0	1	1
		H	0	1	1	0	2	1
H	L	L	0	0	0	0	0	1
		M	0	0	0	0	0	0
		H	0	0	0	0	0	0
	M	L	0	0	0	0	0	0
		M	0	4	4	1	13	0
		H	0	1	4	0	0	0
	H	L	0	0	0	0	0	0
		M	1	2	0	1	7	2
		H	0	5	12	0	2	7

(a) Study how the level of stress changes over time. Note that the interviews were not equally spaced in time.

(b) Is there any difference with distance from the plant?

(10) The table below shows the inter-urban moves of a sample of people over five two-year periods in Milwaukee, Wisconsin, U.S.A. (Crouchley *et al.*, 1982, from Clark *et al.*).

	Renters		Owners	
	Age			
Sequence	25–44	46–64	25–44	46–64
SSSSS	511	573	739	2385
SSSSM	222	125	308	222
SSSMS	146	103	294	232
SSSMM	89	30	87	17
SSMSS	90	77	317	343
SSMSM	43	24	51	22
SSMMS	27	16	62	19
SSMMM	28	6	38	5
SMSSS	52	65	250	250
SMSSM	17	20	48	14
SMSMS	26	19	60	25
SMSMM	8	4	10	3
SMMSS	8	9	54	21
SMMSM	11	3	18	1
SMMMS	10	3	21	1
SMMMM	4	1	8	2
MSSSS	41	29	134	229
MSSSM	16	15	23	10
MSSMS	19	13	36	25
MSSMM	2	4	1	0
MSMSS	11	10	69	24
MSMSM	11	2	15	3
MSMMS	1	9	13	2
MSMMM	2	2	2	0
MMSSS	7	5	40	18
MMSSM	4	2	9	2
MMSMS	8	1	15	3
MMSMM	1	0	5	0
MMMSS	8	1	22	7
MMMSM	3	2	7	2
MMMMS	5	0	9	2
MMMMM	6	3	5	0

(a) Develop a model to describe these data.

(b) Do people tend to move less if they have already stayed a long time in one place?

9
Structured transition matrices

In the previous chapter, we encountered certain models of Markov chains which can be applied to mobility data. There, the models involved square tables, although usually in more than two dimensions. In this chapter, we shall study a number of special models applicable to square tables in particular, such as transition matrices, and we shall concentrate on certain patterns which may occur in such tables.

9.1 Main diagonal models

As we have seen, a mobility table is a square two- (or more) dimensional table with the same categorical variable, defining the states in which a unit can be, observed at two (or more) points in time. If the same subjects are involved, it is a form of panel study. Examples include the British social mobility table of Section 8.1 and the voting change tables of Sections 8.3 and 8.4; the second is a panel study, while the first is a retrospective study.

A series of standard social mobility models have been described by Duncan (1979). These may all easily be fitted with generalized linear modelling software. The most important one, which we shall look at here, involves elimination of specific cells from the table for theoretical reasons, yielding an incomplete table, as in Chapter 5. We shall look at the others below in Section 9.3.

We usually first wish to check if position at the second point in time actually depends on that at the first point in time, whether it be profession, vote, place of residence, or whatever. This is the standard model of independence which we have encountered many times. However, here, as we have already noted with such tables, the problem is that too many individuals do not change position between the two time points for such independence to be acceptable. Too many observations appear on the diagonal. The simple solution is to eliminate these diagonal elements. We can then check for quasi-independence.

More theoretically, this approach assumes that the diagonal contains a mixture of two types of individual, the movers, who might have moved, but did not happen to, in the observed time interval, and the stayers, who never change. Hence, the name of the model: the mover-stayer model. The estimation procedure for such a mixture is the same as that which we used

Table 9.1. Voting changes between the British elections of 1964 and 1966. (Upton, 1978, p.111)

	1964			
1966	Conservative	Liberal	Labour	Abstention
Conservative	157	4	17	9
Liberal	16	159	13	9
Labour	11	9	51	1
Abstention	18	12	11	15

Table 9.2. Fitted values for three models for the British elections data of Table 9.1, with estimates of stayers underlined.

1964	1966	Observed	Independence	Mover-stayer	Loyalty
Cons	Cons	157	73.8	<u>16.8</u>	154.9
Lib		4	67.2	9.4	14.9
Lab		17	33.6	13.4	13.2
Abs		9	12.4	7.2	4.0
Cons	Lib	16	77.7	17.1	24.1
Lib		159	70.8	<u>9.5</u>	151.2
Lab		13	35.4	13.7	16.6
Abs		9	13.1	7.3	5.8
Cons	Lab	11	28.4	10.6	9.6
Lib		9	25.9	5.9	7.4
Lab		51	12.9	<u>8.5</u>	53.0
Abs		1	4.8	4.5	2.0
Cons	Abs	18	22.1	17.4	13.4
Lib		12	20.1	9.7	10.4
Lab		11	10.1	13.9	9.2
Abs		15	3.7	<u>7.4</u>	22.9

for Guttman scales in Section 5.5 and for mixture distributions in Section 6.6: we weight out the diagonal cells.

We apply these models to data for the voting changes between the British elections of 1964 and 1966, given in Table 9.1. The independence model

```
VOTE64 + VOTE66
```

has a deviance of 480.45 (494.45) with 9 d.f., clearly indicating that it is unacceptable. The fitted values, in Table 9.2, show how the diagonal values are underestimated.

We now give zero weight to the diagonal elements and refit the (now quasi-) independence model. With a deviance of 12.33 (34.32) and 5 d.f., this mover-stayer model provides a big improvement, although it is still not a satisfactory fit. For the movers, those individuals who are susceptible to change, the new party chosen does not depend very much on the original party. The fitted values for the diagonal, given by the software and shown in Table 9.2 are, here, prediction estimates of the numbers of movers in each category who did not happen to move in the period under observation. (In fact, the true fitted values on the diagonal for this model are just the observed values, because these observations fit perfectly.) The estimated number of stayers (140.2, 149.5, 42.5, 7.6), is obtained by subtracting these values from the observed diagonal values. We see that 66.4% (339.8/512) of the voters are estimated as not being prepared to change parties.

In fact, this model is only one special case of the mover-stayer model, that where the transition matrix for the movers has independence between the states at the beginning and end of the time period. Other structures for this matrix are also possible, as we shall see. On the other hand, the transition matrix for the stayers is always a diagonal matrix.

The independence mover-stayer model assumes a different probability of staying for each party. Another model bears similarity to this mover-stayer model. We differentiate those who do not change from those who do, i.e. the diagonal from the rest. However, in distinction to the previous model, we here consider the diagonal members to be homogeneous, i.e. all to have the same probability of remaining loyal to the party. This is the main diagonal or loyalty model, because, in voting behaviour, where it was developed, we are distinguishing those who are loyal to a party (those on the diagonal) from those who are not. The factor variable (`LOYAL`) is

2	1	1	1	1	1
1	2	1	1	1	1
1	1	2	1	1	1
1	1	1	2	1	1
1	1	1	1	2	1
1	1	1	1	1	2

The model is now

$$\texttt{VOTE64} + \texttt{VOTE66} + \texttt{LOYAL}$$

but where the diagonal is not weighted out. We may expect that this model will necessarily fit less well than the independence mover-stayer one.

One possible interpretation of the loyalty model is as another special case of the mover-stayer model. Here, the movers still have an independence transition matrix, but the total probability of movers and stayers staying in a period is identical for all states.

Although the deviance of 29.84 (45.84) with 8 d.f. is greatly reduced from that for the independence model, the model is considerably worse than

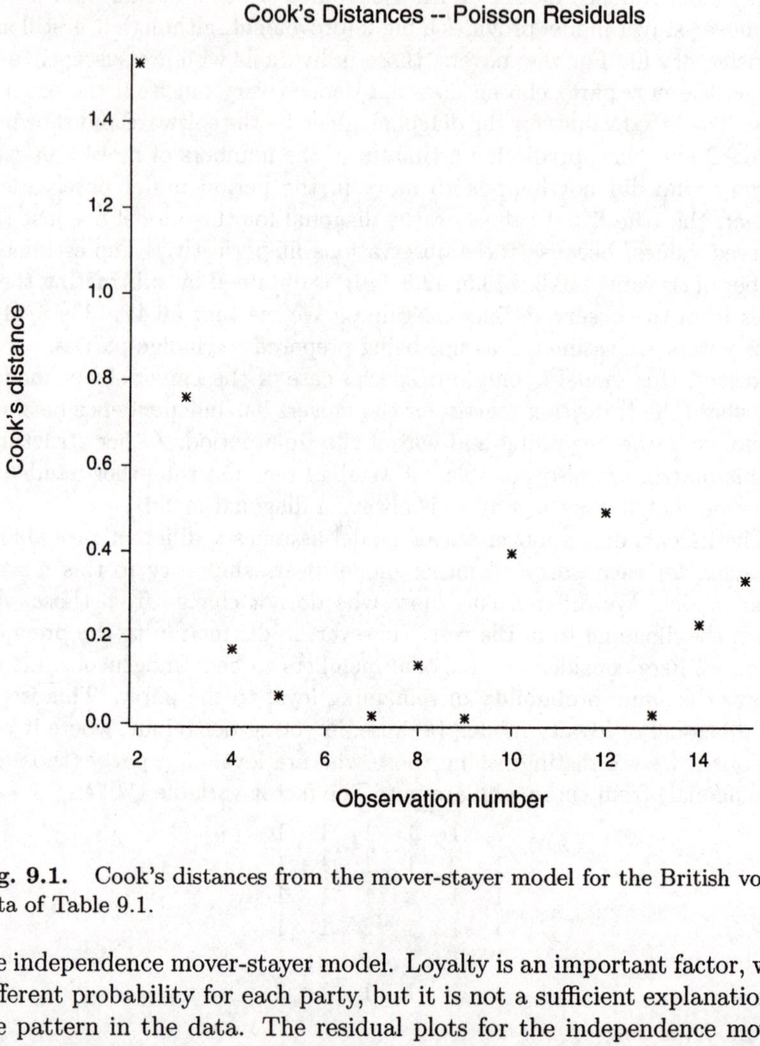

Fig. 9.1. Cook's distances from the mover-stayer model for the British voting data of Table 9.1.

the independence mover-stayer model. Loyalty is an important factor, with different probability for each party, but it is not a sufficient explanation of the pattern in the data. The residual plots for the independence mover-stayer model are given in Figures 9.1 and 9.2. Cook's distances indicate that moves of Liberal and Labour to the Conservative party (observations 2 and 3) have the worst fit, while the probability plot shows a problem with the negative residuals (observations 2 and 12). One of the models in the following sections will certainly fit these data better.

9.2 Symmetry models

A completely symmetrical table is one in which the probabilities in opposing cells across the diagonal are equal:

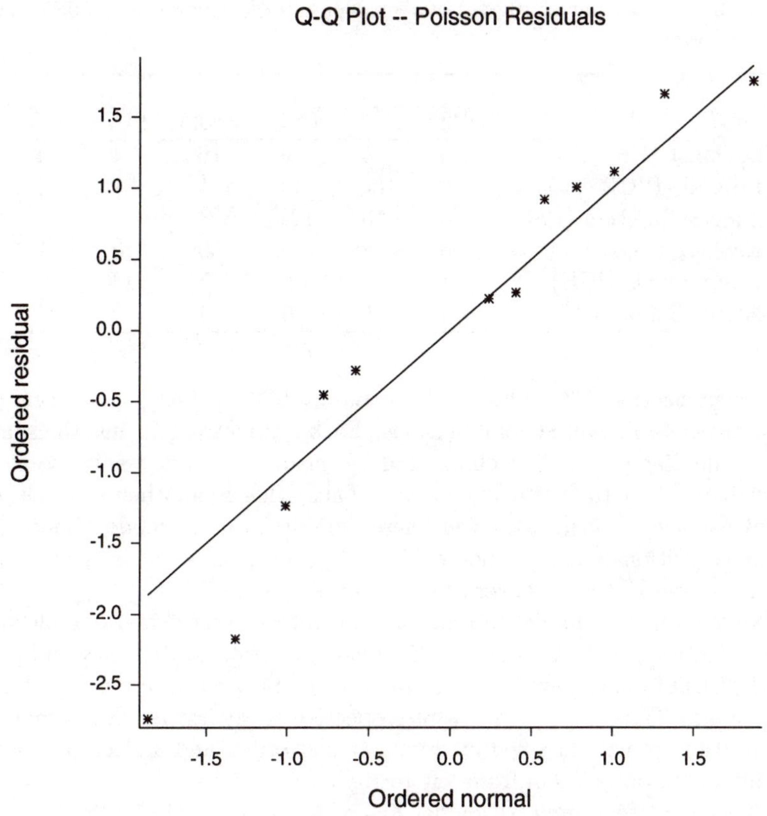

Fig. 9.2. The normal probability plot from the mover-stayer model for the British voting data of Table 9.1.

$$\pi_{ij} = \pi_{ji} \tag{9.1}$$

The corresponding log linear model is

$$\log(\nu_{ij}) = \alpha_{ij} \qquad \text{with } \alpha_{ij} = \alpha_{ji} \tag{9.2}$$

Notice that mean parameters for the margins are not fitted. We construct a symmetric factor variable in the following way, here with six categories:

—	1	2	3	4	5
1	—	6	7	8	9
2	6	—	10	11	12
3	7	10	—	13	14
4	8	11	13	—	15
5	9	12	14	15	—

Table 9.3. Voting changes between Belgian elections, 1981–1985. (R. Doutrelepont)

	1981					
1985	PS	PRL	PSC	Ecolo	PCB	BB
Socialist (PS)	281	14	9	16	4	4
Liberal (PRL)	12	164	13	4	1	6
Social-Christian (PSC)	5	10	121	8	1	1
Ecology (Ecolo)	6	0	1	50	0	1
Communist (PCB)	1	0	0	2	14	0
Blank Ballot (BB)	2	1	0	0	0	11

just as in Section 5.3. This factor variable, `SYM`, has as many levels as there are possible paired combinations. Each symmetric pair has the same level. The diagonal will be eliminated by means of a zero weight, so that the values given to it are irrelevant. Thus, this is another example of a mover-stayer model. However, here, the assumption made about the mover transition matrix is not sufficiently strong to allow estimation of stayers with observations over only one period.

We shall fit this model to data on how voters, interviewed outside the polling stations in the October, 1985, Belgian election, stated they had just voted (`VOTE85`) and how they had voted in the previous election (`VOTE81`), as shown in Table 9.3. This is a retrospective study, rather than a panel. The parties are not classified in any particular order, and, indeed, it would be difficult to order them from left to right.

The complete symmetry model has a deviance of 33.05 (75.05) with 15 d.f. Because this model is rejected, the probability of changing vote in either direction between each pair of parties is not the same.

A weaker hypothesis is that of quasi-symmetry: the table would be symmetric if it were not for the distorting effect of the marginal totals. In our example, this is the effect of a changing proportion of votes received by the different parties between the two elections. This is another mover-stayer model. We simply add the two marginal variables to the model:

$$\texttt{VOTE81} + \texttt{VOTE85} + \texttt{SYM} \tag{9.3}$$

With a deviance of 10.10 (62.10) and 10 d.f., this model fits the data well. The probability of shifting in either direction between each pair of parties is the same after taking into account the overall change in voting behaviour between the two elections. The parameter estimates for `SYM`, given in Table 9.4, provide us with the required information. Changes between Socialist and Liberal are most probable (all other estimates are negative with regard to it) and between Communist and blank ballot least

Table 9.4. Parameter estimates from the quasi-symmetry model of Equation (9.3) for the Belgian election data of Table 9.3.

		Estimate	s.e.
PS	PRL	0.000	—
	PSC	−0.744	0.376
	Ecolo	−1.189	0.455
	PCB	−2.353	0.751
	BB	−2.510	0.707
PRL	PSC	−0.228	0.345
	Ecolo	−2.886	0.638
	PCB	−3.951	1.167
	BB	−2.348	0.693
PSC	Ecolo	−2.122	0.521
	PCB	−4.016	1.158
	BB	−4.340	1.148
Ecolo	PCB	−3.894	0.849
	BB	−4.783	1.097
PCB	BB	−13.59	54.27

(there are none observed). These two sampling zeros also explain the large standard error for the latter estimate.

A further model, marginal homogeneity, is closely related to the previous two. Suppose that the marginal totals are symmetric but the body of the table is not. The distribution of votes at the two elections is identical, but the probability of shift between each pair of parties is not the same in both directions. Marginal homogeneity plus quasi-symmetry equals symmetry; the deviances of these models obey this equation. Symmetry obviously implies marginal homogeneity. As might be expected, with a deviance of 22.95 (84.95) and 5 d.f., this model is not acceptable for these data, because symmetry was not, while quasi-symmetry was.

We may note that marginal homogeneity is not a log linear model; this is the third such model which we have encountered in this book; the first two were the log multiplicative and proportional odds models. (A GLIM4 macro for this model is given in Appendix A.3.)

In terms of Markov chains, quasi-symmetry is known as reversibility, because the same proportion of individuals is changing position in each direction, while marginal homogeneity is the equilibrium state, because the margins are not changing over time. This is stationarity of the marginal distribution of states, as opposed to the stationarity of the transition matrix, studied in Section 8.3.

Table 9.5. Migrant behaviour in Britain between 1966 and 1971. (Fingleton, 1984, p. 142)

	1971			
1966	Central Clydes.	Urban Lancs. and Yorks.	West Midlands	Greater London
Central Clydesdale	118	12	7	23
Urban Lancs. and Yorks.	14	2127	86	130
West Midlands	8	69	2548	107
Greater London	12	110	88	7712

9.3 Ordered categories: random walks

In Section 9.1, we looked at one of Duncan's (1979) standard mobility models, the independence mover-stayer model. For the other models which he proposes, we must assume an ordering for the categories and fit an equal interval scale. In this context, let us look at data on migrant behaviour, given in Table 9.5. We note, in this migration table, that the geographical locations are ordered from the north to the south of Britain. We also observe the exceptionally high values on the diagonal.

The first model fitted is the usual one for independence

$$\texttt{RES66} + \texttt{RES71} \tag{9.4}$$

This has a deviance of 19 884 (19 898) with 9 d.f. As would be expected, this model is definitely not acceptable. In column four of Table 9.6, we see the large underestimation of all diagonal cells. Let us note, in passing, that the independence mover-stayer model of Section 9.1 has a deviance of 4.37 (26.37) with 5 d.f., a very good fit.

The second Duncan model, called row effects (assuming that the first time point forms the rows of the table), takes the second position as a linear equal interval scale (`LRES71`) and the first position as a nominal variable and fits the interaction between them,

$$\texttt{RES66} + \texttt{RES71} + \texttt{RES66} \cdot \texttt{LRES71} \tag{9.5}$$

Although a major improvement, with a deviance of 4155.6 (4175.6) and 6 d.f., the model is still not acceptable. The diagonal estimates are, however, better, as seen in column five of Table 9.6. The parameter estimates, in Table 9.7, are the slopes for each category of origin, each calculated in relation to the last category. As can also be seen from the table, all slopes are negative relative to the last line of the table. However, migration both ways between Clydesdale and London is especially underestimated. This is due to the linear scale which should continue to decrease from Clydesdale,

Table 9.6. Fitted values from two models, (9.4) and (9.5), for the migrant data of Table 9.5.

1966	1971	Observed	Independence	Row effect
CC	CC	118	1.8	66.3
	ULY	12	28.2	92.4
	WM	7	33.2	1.3
	GL	23	96.8	0.0
ULY	CC	14	27.2	85.0
	ULY	2127	414.8	1858.3
	WM	86	488.4	410.4
	GL	130	1426.6	3.3
WM	CC	8	31.5	0.7
	ULY	69	480.8	366.7
	WM	2548	566.1	1974.5
	GL	107	1653.6	390.1
GL	CC	12	91.4	0.0
	ULY	110	1394.2	0.6
	WM	88	1641.4	342.9
	GL	7712	4794.9	7578.6

Table 9.7. Parameter estimates from the row effect model of Equation (9.5) for the migrant data of Table 8.4.

	With diagonal		Without diagonal	
	Estimate	s.e.	Estimate	s.e.
Central Clydesdale	−10.66	0.219	0.171	0.249
Urban Lancs. and Yorks.	−7.911	0.113	−0.144	0.200
West Midlands	−4.717	0.078	0.024	0.171
Greater London	0.000	—	0.000	—

through Lancashire, Yorkshire, and the West Midlands to London, but increases for London in the first row of Table 9.5.

Because the remaining two models are generalizations of the independence mover-stayer model, with additional parameters, we may expect that they will provide acceptable fits. The first of these assumes both variables to have equal interval scales and fits the interaction between them

$$\texttt{RES66} + \texttt{RES71} + \texttt{LRES66} \cdot \texttt{LRES71}$$

It has a deviance of 4.36 (28.36) with 4 d.f. The parameter estimate, 0.0049, gives the slope of the relationship between the two scales. Here, the almost zero slope reflects the independence of new residence from place of origin for the movers.

Finally, the row effects model is refitted, but now without the diagonal. The deviance is 1.51 (29.51) with 2 d.f. When the diagonal is eliminated, the slopes are greatly reduced as compared to the second model above.

In both of these last models, the fit is very good, but too many parameters are included in the model. The simpler independence mover-stayer model is retained as that best describing the data. The most important conclusions are the small proportion of movers in the population and the independence of arrival point from origin for these movers.

Several models have been developed to take into account the diagonal symmetry of square tables with ordered categories. We now take into consideration, in another way, the fact that the locations in Britain are ordered and that changing location by one step in either direction on the scale has a different probability than that for two steps, and so on, for greater distances. The factor variable (SDIAG) is now

1	2	3	4	5	6
2	1	2	3	4	5
3	2	1	2	3	4
4	3	2	1	2	3
5	4	3	2	1	2
6	5	4	3	2	1

the symmetric minor diagonal model. In terms of Markov chains, this describes a random walk, without drift, because probabilities are equal in both directions, where jumps of more than one step are possible. This model,

$$\texttt{RES66} + \texttt{RES71} + \texttt{SDIAG}$$

fits the data poorly, with a deviance of 26.84 (46.84) and 6 d.f. A special case of this model is the more usual random walk, with jumps of only one step, for which all minor diagonals, except the one on each side of the main diagonal, would be zero.

Because the model did not fit well, we can extend it further and take steps with different probabilities in each direction, the asymmetric minor diagonal model, with factor variable (ADIAG)

1	2	3	4	5	6
7	1	2	3	4	5
8	7	1	2	3	4
9	8	7	1	2	3
10	9	8	7	1	2
11	10	9	8	7	1

This is a random walk with drift. Unfortunately, this model

Table 9.8. Voting changes between 1966 and 1970 British elections. (Upton, 1978, p. 119)

	1970			
1966	Conservative	Liberal	Labour	Abstention
Conservative	68	1	1	7
Liberal	12	60	5	10
Labour	12	3	13	2
Abstention	8	2	3	6

$$\texttt{RES66} + \texttt{RES71} + \texttt{ADIAG}$$

also fits poorly, with a deviance of 24.07 (48.07) and 4 d.f. From what we saw above, the problem is that these two models do not take into account the large diagonal values.

Still another possibility is to combine the minor diagonals model with symmetry. When this model fits, it indicates that we would have symmetry if it were not for the unequal probabilities of larger and smaller steps. We fit it with the symmetry factor variable, but with no mean parameters:

$$\texttt{SDIAG} + \texttt{SYM}$$

This model is similar to the symmetric minor diagonal model, but without the margins fixed. With a deviance of 23.71 (49.71) and 3 d.f., the fit is still poor. Thus, for these migration data, we are left with the independence mover-stayer model as preferable.

9.4 Ordered categories: distance models

Minor diagonal models assume an equal distance among all adjacent pairs of ordered categories. Distance models relax this assumption to allow different intervals among the categories. A distinct variable is introduced for each adjacent interval, i.e. $I-1$ variables for I categories. A model with these variables plus the two mean variables may be called a pure distance model. For a 4×4 table, the series of variables (D1, D2, D3) is

$$\begin{matrix} 2 & 1 & 1 & 1 \\ 1 & 2 & 2 & 2 \\ 1 & 2 & 2 & 2 \\ 1 & 2 & 2 & 2 \end{matrix} \qquad \begin{matrix} 2 & 2 & 1 & 1 \\ 2 & 2 & 1 & 1 \\ 1 & 1 & 2 & 2 \\ 1 & 1 & 2 & 2 \end{matrix} \qquad \begin{matrix} 2 & 2 & 2 & 1 \\ 2 & 2 & 2 & 1 \\ 2 & 2 & 2 & 1 \\ 1 & 1 & 1 & 2 \end{matrix}$$

We fit the model to data on voting changes between the 1966 and 1970 British elections in Table 9.8. The model is

$$\texttt{VOTE66} + \texttt{VOTE70} + \texttt{D1} + \texttt{D2} + \texttt{D3} \tag{9.6}$$

Table 9.9. Fitted values and residuals for the pure distance model of Equation (9.6) applied to the British voting data of Table 9.8.

1966	1970	Observed	Fitted	Residual
Cons	Cons	68	68.0	0.000
	Lib	1	6.2	−2.077
	Lab	1	1.4	−0.329
	Abs	7	1.5	4.588
Lib	Cons	12	23.3	−2.339
	Lib	60	43.6	2.491
	Lab	5	9.8	−1.540
	Abs	10	10.3	−0.101
Lab	Cons	12	5.5	2.741
	Lib	3	10.4	−2.289
	Lab	13	6.9	2.341
	Abs	2	7.2	−1.942
Abs	Cons	8	3.2	2.718
	Lib	2	5.9	−1.611
	Lab	3	3.9	−0.464
	Abs	6	6.0	0.000

With a deviance of 64.23 (84.23) and 6 d.f., the model is rejected. Again, the problem with such a model for voting data is that it does not take into account party loyalty. If we inspect the residuals in Table 9.9 and Cook's distances (not shown), we see that this is only important for the interior diagonal elements (Liberal and Labour), and not for the two extremes. The stability of the Liberal vote is particularly underestimated.

We add the loyalty variable (`LOYAL`), the main diagonal factor of the previous section, to obtain the distance loyalty model:

$$\texttt{VOTE66} + \texttt{VOTE70} + \texttt{D1} + \texttt{D2} + \texttt{D3} + \texttt{LOYAL} \tag{9.7}$$

The model now fits very well, with a deviance of 6.07 (28.07) and 5 d.f. (In fact, the symmetric minor diagonal model also fits these data well; it too takes party loyalty into account.) On the ordered party scale, Conservative and Liberal are closest neighbours (0.064) and Labour and abstention are most distant (0.440), with the Liberal–Labour distance (0.315) in between.

Another possibility for accommodating the inflated main diagonal elements is to generalize the pure distance model to a mover-stayer model, giving a distance mover-stayer model. This model also fits the data well, having a deviance of 4.30 (28.30), but with one less degree of freedom than the previous model. As we have seen, whereas the main diagonal variable gives a constant factor level, and constant probability, to the whole diagonal, eliminating the diagonal is equivalent to giving each category of the

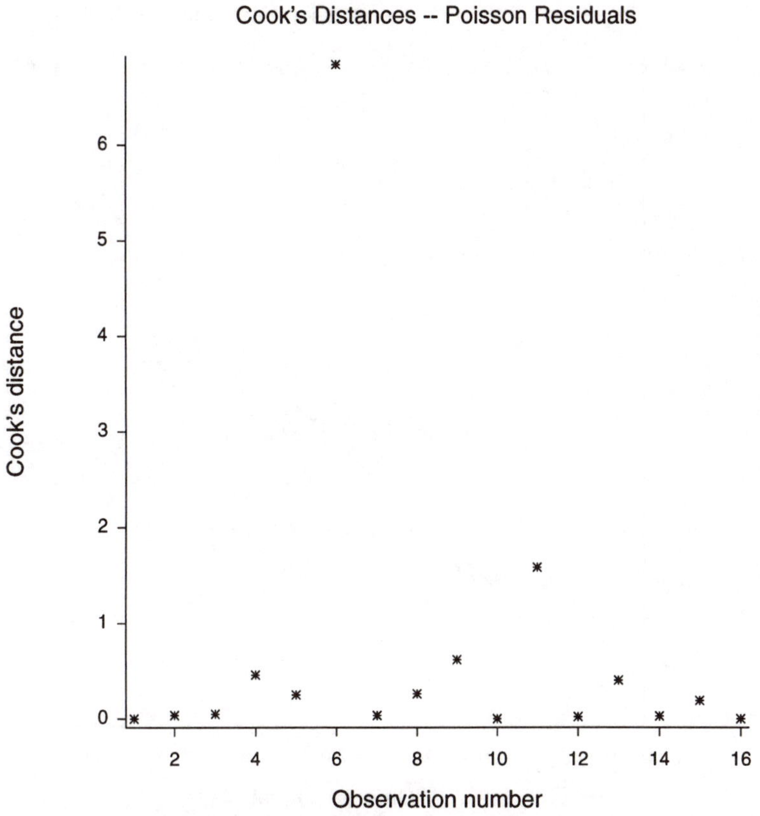

Fig. 9.3. Cook's distances from the distance loyalty model of Equation (9.7) for the British voting data of Table 9.8.

diagonal a different level. This is evidently unnecessary for these data.

The residual plots for the distance loyalty model are given in Figures 9.3 and 9.4. Only the loyalty of Liberals between the two elections appears to pose some problem, as seen by its Cook's distance.

Several remarks should be made in conclusion. All of the models of this chapter are suitable for the analysis of transition matrices, already introduced in the previous chapter. However, care must be taken with small tables, especially 3×3, because, in this case, a number of the models are identical. Note, also, that all of these models may be relatively easily extended to multi-dimensional tables covering more than two time points.

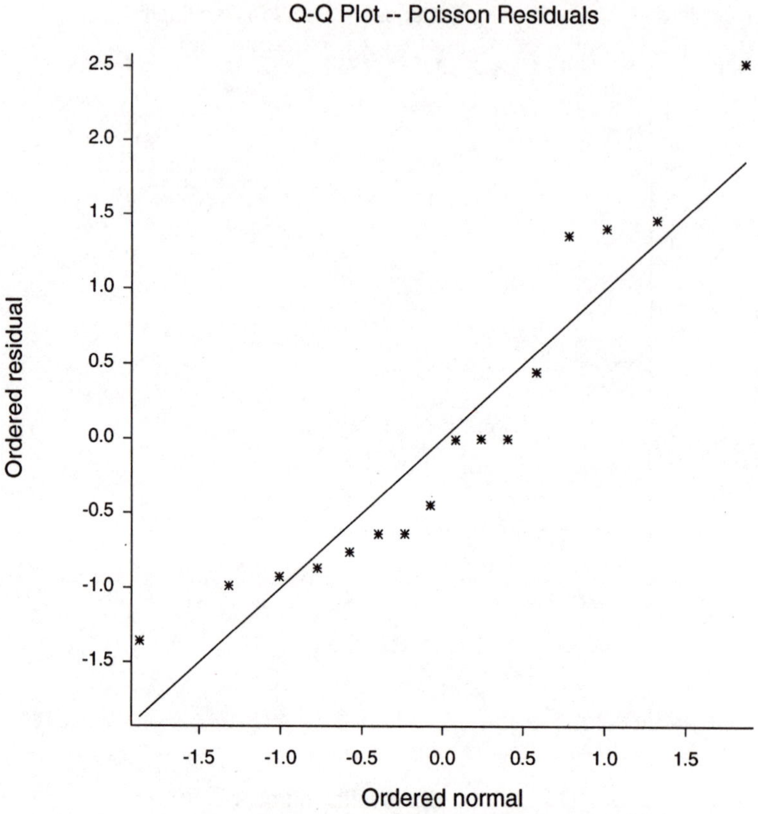

Fig. 9.4. The normal probability plot from the distance loyalty model of Equation (9.7) for the British voting data of Table 9.8.

9.5 Mover-stayer models

In the previous sections of this chapter, we have studied a number of possible versions of the mover-stayer model. It is characterized by a mixture of two types of individual on the main diagonal for each transition between periods. The stayers have a diagonal transition matrix because they never move. The movers can have any arbitrary transition matrix, depending on the model. In this section, we shall look briefly at such models when data are available for more than two periods. As time passes, more and more information accumulates, especially about who are the stayers.

Consider Table 9.10, which shows the labour force participation of married women in the U.S.A. between 1967 and 1971. Here, there are two possible types of stayer: the women who work all the time and those who are out of work all the time.

Table 9.10. Labour force participation of married women. (Heckman and Willis, 1977)

				1971	
1970	1969	1968	1967	Yes	No
Yes	Yes	Yes	Yes	426	38
No				16	47
Yes	No			11	2
No				12	28
Yes	Yes	No		21	7
No				0	9
Yes	No			8	3
No				5	43
Yes	Yes	Yes	No	73	11
No				7	17
Yes	No			9	3
No				5	24
Yes	Yes	No		54	16
No				6	28
Yes	No			36	24
No				35	559

Let us first attempt to determine the order of the Markov chain required for these data. The independence model is

$$\texttt{WORK71} + \texttt{WORK70} + \texttt{WORK69} + \texttt{WORK68} + \texttt{WORK67}$$

with a deviance of 3824.9 (3836.9) and 26 d.f. The first-order chain has

$$\texttt{WORK71} * \texttt{WORK70} + \texttt{WORK70} * \texttt{WORK69} + +\texttt{WORK69} * \texttt{WORK68} \\ +\texttt{WORK68} * \texttt{WORK67}$$

with deviance 210.23 (230.23) and 22 d.f., still not a good fit. When we move to second order

$$\texttt{WORK71} * \texttt{WORK70} + \texttt{WORK71} * \texttt{WOKR69} + \texttt{WORK70} * \texttt{WORK69} + \texttt{WORK70} * \texttt{WORK68} \\ +\texttt{WORK69} * \texttt{WORK68} + \texttt{WORK69} * \texttt{WORK67} + \texttt{WORK68} * \texttt{WORK67}$$

the deviance is 63.22 (89.22) with 19 d.f. Finally, with a third order chain,

$$\texttt{WORK71} * \texttt{WORK70} + \texttt{WORK71} * \texttt{WORK69} + \texttt{WORK71} * \texttt{WORK68} + \texttt{WORK70} * \texttt{WORK69} \\ +\texttt{WORK70} * \texttt{WORK68} + \texttt{WORK70} * \texttt{WORK67} + \texttt{WORK69} * \texttt{WORK68} \\ +\texttt{WORK69} * \texttt{WORK67} + \texttt{WORK68} * \texttt{WORK67}$$

Table 9.11. Deviances (AIC), and degrees of freedom, for various orders of Markov chains for the women working data of Table 9.10.

	Without		With	
Order	Mover-stayer			
0	3824.9 (3836.9)	26	394.91 (410.91)	24
1	210.23 (230.23)	22	36.71 (60.71)	20
2	63.22 (89.22)	19	22.03 (54.03)	17
3	16.19 (46.19)	17	8.49 (42.49)	15

we have a satisfactory deviance of 16.19 (46.19) with 17 d.f. Notice that we have not used higher order interactions, such as `WORK71*WORK70*WORK69`; nor are they necessary. And only the dependency of working in 1970 on working in 1968 could be removed.

We have not yet used any mover-stayer assumption. However, inspection of the table reveals large frequencies of women who were working all five years and of those not working all five years. If we weight these two out, the corresponding deviances for the above models are summarized in Table 9.11. The introduction of stayers improves the model, although it may be overfitting when we reach the required order. The estimated numbers of stayers are 186 working and 268 unemployed, or, respectively, 1.2% and 1.7% of the population. We might believe that this was because there are only stayers of one type. However, if we only allow unemployed stayers, the deviance rises to 14.12 (46.12) with 16 d.f.

9.6 Exercises

(1) The table below gives initial and followup blood pressure according to hypertension status (Lawal and Upton, 1990, from Freeman).

	Followup		
Initial	Normal	Borderline	Elevated
Normal	105	9	3
Borderline	10	12	1
Elevated	3	2	7

Find an appropriate model to describe these data.

(2) The table in Exercise (6.15) gave the numbers of lambs born to ewes in two successive years. Try to find an appropriate transition matrix model to describe these data.

(3) The table below shows the changes in vote between the two British elections of 1974 (Fingleton, 1984, p. 131).

February	October		
	Conservative	Liberal	Labour
Conservative	170	20	3
Liberal	22	70	28
Labour	6	12	227

(a) Find an appropriate model to describe these data.

(b) Which of the models of this chapter give the same fit for these data? Why?

(4) Subjects in the General Social Survey in the U.S.A. were classified by their residence at the time of the survey in 1974 and that at age 16 (Haberman, 1979, p. 445).

Age 16	Current residence			
	Northeast	South	Northcentral	West
Northeast	263	33	14	13
South	26	399	36	30
Northcentral	10	41	368	46
West	1	8	5	148

Find an appropriate model to describe these data.

(5) The following table concerns Danish social mobility (Bishop *et al.*, 1975, p. 100, from Svalastoga).

Father	Son				
	1	2	3	4	5
1	18	17	16	4	2
2	24	105	109	59	21
3	23	84	289	217	95
4	8	49	175	348	198
5	6	8	69	201	246

1 — Professional and high administrative
2 — Managerial, executive and high supervisory
3 — Low inspectional and supervisory
4 — Routine non-manual and skilled manual
5 — Semi- and unskilled manual

Find an appropriate model to describe these data.

(6) The table below is from a study of voting during the Swedish elections of 1964 and 1970 (Fingleton, 1984, p.138).

1964	1970				
	Comm.	SD	Centre	People's	Cons.
Communist	22	27	4	1	0
Social Dem.	16	861	57	30	8
Centre	4	26	248	14	7
People's	8	20	61	201	11
Conservative	0	4	31	32	140

Find an appropriate model to describe these data.

(7) (a) Find suitable mover-stayer models to fit
   i. the nuclear accident data of Exercise (8.9);
   ii. the inter-urban move data of Exercise (8.10).

   (b) In each case, does the new model change your conclusions substantially from those of Chapter 8?

(8) Frequency of company moves between British regions was recorded between 1972 and 1978 (Upton and Fingleton, 1989, p. 6):

	1978									
1972	SE	EA	SW	WM	EM	Y	NW	N	W	S
SE	958	119	70	16	95	48	39	46	89	41
EA	10	121	44	0	7	2	8	8	3	6
SW	9	11	162	5	2	2	9	6	25	6
WM	7	0	1	288	13	4	10	7	33	6
EM	10	9	5	6	332	16	4	6	6	4
Y	6	1	2	1	20	356	3	9	1	9
NW	8	1	10	6	10	15	331	12	21	8
N	0	2	4	1	5	4	5	119	1	3
W	5	0	3	1	0	0	0	2	154	1
S	1	1	2	0	1	2	4	5	4	269

SE — South East EA — East Anglia SW — South West
WM — West Midlands EM — East Midlands Y — Yorkshire
NW — North West N — North W — Wales
S — Scotland

Find an appropriate model to describe these data.

(9) The table in Exercise (8.7) gave the two-step transitions for Swedish voting between 1964 and 1970.

   (a) Develop a structured transition matrix model to describe these data.

   (b) Is it superior to what was found in Chapter 8?

   (c) How does it compare with the results found above in Exercise (9.6)?

(10) When we studied the order of the Markov chain for the data in Table 8.6, we noted the poor fit of the three cells for complete stability of voter intention. Check order again, using a mover-stayer model.

(11) The following table gives the inter-provincial migration (five years old and over) in Canada between 1976 and 1981 (Upton and Fingleton, 1989, p. 149, from Ledent).

Origin	Nfld	PEI	Nova Scotia	NB	Quebec
	Destination				
Nfld	—	556	6089	2164	1420
PEI	250	—	2370	1383	181
NS	3130	1952	—	8467	2997
NB	1311	1604	8927	—	6372
Que.	1471	923	6146	10188	—
Ont.	9719	3309	20933	10004	39382
Man.	717	118	1898	1033	1793
Sask.	178	228	807	513	776
Alb.	766	655	3591	1831	2988
BC	590	511	3369	1734	4739
	Ontario	Manitoba	Sask.	Alberta	BC
Nfld	13263	1443	704	8836	3092
PEI	2255	187	215	2362	671
NS	20062	1962	1494	14585	7617
NB	13728	1426	925	11187	4144
Que.	123708	4263	2386	31296	21769
Ont.	—	21743	13150	126746	75834
Man.	17853	—	13532	34642	24882
Sask.	6770	7337	—	35028	16670
Alb.	24672	8114	19855	—	73857
BC	26661	7030	10143	65408	—

Find an appropriate model to describe these data.

10
Overdispersion and cluster models

As we have seen in previous chapters of this part, count data arise from the enumeration of events on individual units. If no explanatory variables, or time, distinguish among the events, they may be aggregated as counts. In fact, this can only be done legitimately for Poisson or binomial data if the events are independent. One may expect that such would not be the case for repeated events on each unit. This is usually known as over- or underdispersion, depending on whether dependence is positive or negative.

Over- (or under-) dispersion in counts can arise for at least two reasons.

(1) The probability of an event occurring to a unit may be the same for all units, but depend, in some way, on the previous events happening to that unit. In other words, it varies over time. This might, for example, be a contagion model, whereby the occurrence and timing of an event may either increase or decrease the subsequent probability of events for that unit.

(2) The probability of an event may remain constant over time but not necessarily be the same for all units. This is the heterogeneous population model of varying susceptibility. We may distinguish two cases:
   (a) If causes are internal, we have frailty or proneness.
   (b) If they are external, for example due to environmental differences, we have liability.

These two models are not distinguishable from a given data set containing only aggregated counts. The first was covered, for disaggregated data, in the previous chapters of this part. We shall look at the second here.

10.1 Overdispersion

One simple way to correct for overdispersion in aggregated counts is to modify the fixed relationship between the mean and variance in the Poisson or binomial distribution by including a proportionality constant, called the heterogeneity factor. Thus, for the Poisson distribution, we would have $\text{var}[Y] = \phi\nu$ and, for the binomial distribution, $\text{var}[Y] = n_{\bullet}\phi\pi(1-\pi)$. This factor, ϕ, can be estimated quite simply from the deviance of the most complex model considered, provided that it contains all appropriate explanatory variables. It is obtained by dividing the deviance by the degrees of freedom, although this is not the maximum likelihood estimate, but a moment estimate.

As an example of overdispersion in binomial count data, consider data on the proportion of Pacific cod (*Gadus macrocephalus*) eggs hatching in sets of four tanks under a number of different conditions of salinity, temperature, and dissolved oxygen, as given in Table 10.1. This is a response surface experiment in which the optimum combination of these three factors for the fertility of the eggs was sought. Each salinity–temperature–oxygen combination was replicated four times in different tanks. The problem is the uncontrollable differences among tanks at the same salinity, temperature, and dissolved oxygen. In spite of the similar responses among the four tanks, the resulting variation is greater than would be expected for a binomial distribution, i.e. for independence among eggs in a tank.

We shall begin by fitting two relatively simple models to these data, a model with a different factor level for each combination of salinity, temperature, and oxygen,

$$\mathtt{SALF} * \mathtt{TEMPF} * \mathtt{OXF}$$

where F indicates a factor variable, and a quadratic response surface model,

$$\begin{aligned}\mathtt{SAL} + \mathtt{TEMP} + \mathtt{OX} + \mathtt{SAL2} + \mathtt{TEMP2} + \mathtt{OX2}\\ +\mathtt{SAL} \cdot \mathtt{TEMP} + \mathtt{SAL} \cdot \mathtt{OX} + \mathtt{TEMP} \cdot \mathtt{OX}\end{aligned} \qquad (10.1)$$

where the 2 indicates the square of a variable. The former is the best possible model which can be fitted with the explanatory variables available. The latter model provides a simpler description of the response surface which can be visualized and interpreted. It is especially useful when the goal of the experiment is to explore optimal conditions for the response.

For the standard binomial distribution, we obtain a deviance of 968.26 (1004.26) with 58 d.f. for the factor level model and 1774.8 (1794.9) with 66 d.f. for the response surface model. Neither model fits very well, although the response surface is considerably worse. The difference in deviance is 806.4 with 8 d.f. Because the models fit so badly, the standard errors, in Table 10.2, are greatly underestimated and the differences in deviance overestimated. The simple solution is to estimate a scale parameter or heterogeneity factor in the variance from the mean deviance of the best binomial model, here the factor model, as described above. For this model, the estimate is 16.69 (968.26/58). This obviously does not change the deviance or the parameter estimates. The standard errors are increased by a factor which is the square root of this scale parameter, as seen in Table 10.2. Now, differences in deviance can no longer be compared directly. However, an *ad hoc* means of calibration is to compare ratios of mean

Table 10.1. Experimental hatching of Pacific cod eggs under different conditions of salinity (ppt), temperature (°C), and dissolved oxygen (ppm). (Lindsey *et al.*, 1970)

Salinity	Temperature	O_2	Hatch	Total	Hatch	Total
14.00	2.70	3.60	224	283	180	245
			160	235	182	320
14.00	2.70	8.60	231	325	237	283
			171	207	178	270
14.00	9.30	3.60	159	240	163	229
			234	349	295	385
14.00	9.30	8.60	186	314	214	297
			97	298	74	244
26.00	2.70	3.60	5	217	5	316
			2	243	3	224
26.00	2.70	8.60	143	292	186	316
			159	301	138	264
26.00	9.30	3.60	19	262	18	263
			36	277	44	290
26.00	9.30	8.60	74	293	152	307
			68	181	45	167
20.00	6.00	6.10	221	259	224	296
			238	277	281	333
12.71	6.00	6.10	222	268	279	341
			197	289	294	350
27.29	6.00	6.10	46	230	62	214
			243	370	138	265
20.00	2.00	6.10	20	230	10	233
			11	175	7	236
20.00	10.00	6.10	130	389	98	247
			119	226	122	292
20.00	6.00	3.08	187	293	214	271
			168	258	179	220
20.00	6.00	9.14	220	247	159	173
			237	259	275	322
20.00	6.00	6.10	268	475	416	518
			272	493	376	426
19.28	5.43	12.40	197	353	344	420
			262	312	282	400
16.77	5.75	9.46	432	472	730	845
			640	784	736	824
14.25	6.05	6.53	320	375	357	425
			301	406	391	473

Table 10.2. Parameter estimates from two quadratic models (10.1) and (10.2) for the egg hatching data of Table 10.1.

	Binomial		Heterogeneity	Beta-binomial	
	Estimate	s.e.	factor s.e.	Estimate	s.e.
1	0.652	0.4159	1.6990	0.682	1.6380
SAL	−0.290	0.0360	0.1470	−0.290	0.1409
TEMP	1.600	0.0433	0.1768	1.678	0.1693
OX	−0.136	0.0471	0.1923	−0.190	0.1872
SAL^2	−0.005	0.0008	0.0034	−0.006	0.0033
TEMP^2	−0.123	0.0027	0.0109	−0.129	0.0107
OX^2	−0.024	0.0023	0.0093	−0.023	0.0096
SAL·TEMP	0.010	0.0013	0.0053	0.012	0.0050
SAL·OX	0.045	0.0018	0.0073	0.048	0.0069
TEMP·OX	−0.048	0.0032	0.0129	−0.052	0.0123

deviances to F distributions. Thus, to compare the factor and response surface models, we have (1774.8 – 968.26)/8/16.69 = 6.04 with 8 and 58 d.f.

Because the response surface model fits much more poorly than the full factor level model and because we are interested in estimating the response *surface*, we proceed now to fit a non-linear regression model. Following Lindsey (1975), we iteratively estimate power transformations of the three explanatory variables, by including an extra, linearized term for each non-linear parameter. (A GLIM4 macro is given in Appendix A.4.) The deviance for the binomial model is reduced to 1158.2 (1184.2) with 63 d.f., still considerably larger than the deviance of the factor level model. The estimated power transformations are −1.37 for the temperature, −0.58 for the salinity, and 1.92 for the oxygen. This gives the model

$$\begin{aligned}\log\left[\frac{\pi_i}{1-\pi_i}\right] &= -18.45 + 685.9x_{1i}^{-1.37} + 53.55x_{2i}^{-0.58} + 0.059x_{3i}^{1.92} \\ &\quad -13352x_{1i}^{-2.74} - 71.22x_{2i}^{-1.16} - 0.00029x_{3i}^{3.85} \\ &\quad +161.3x_{1i}^{-1.37}x_{2i}^{-0.58} - 3.164x_{1i}^{-1.37}x_{3i}^{1.92} \\ &\quad +0.100x_{2i}^{-0.58}x_{3i}^{1.92} \qquad (10.2)\end{aligned}$$

As compared to the factor model, we have (1158.2−968.26)/5/16.69 = 2.28 with 5 and 58 d.f. This model seems to indicate that we should be using the reciprocal of salinity, the reciprocal square root of temperature, and the square of oxygen, although the latter model has a deviance of 1312.2.

This final response surface equation can be used to study the optimal conditions for egg hatch, although it is difficult to visualize a four-dimensional surface.

Table 10.3. Number of faults in 32 rolls of fabric of various lengths. (Bissell, 1972)

Length	Faults	Length	Faults	Length	Faults
551	6	441	8	657	9
651	4	895	28	170	4
832	17	458	4	738	9
375	9	642	10	371	14
715	14	492	4	735	17
868	8	543	8	749	10
271	5	842	9	495	7
630	7	905	23	716	3
491	7	542	9	952	9
372	7	522	6	417	2
645	6	122	1		

10.2 Compound distributions

A more sophisticated way in which to handle overdispersion is to model it directly. Suppose that each individual unit has a different (Poisson or binomial) parameter, and that this parameter has some distribution, say $p(\cdot)$, over the population. The (marginal) distribution of the observations can be obtained as a compound distribution,

$$\Pr(y_{i1} \ldots y_{iR}) = \int_{\Lambda} \left[\prod_{k=1}^{R} \Pr(y_{ik}|\lambda_i) \right] p(\lambda_i) d\lambda_i \tag{10.3}$$

where i indexes units, k indexes repeated responses on a unit, and λ is either ν for the Poisson or π for the binomial distribution. For aggregated count data, $R = 1$. Suitable choice of $p(\cdot)$ is difficult. One possibility is what is called the conjugate distribution. This yields particularly simple results, because the integral in Equation (10.3) can be solved analytically. The conjugate distribution for the Poisson is the gamma, yielding a negative binomial distribution. That for the binomial is the beta distribution, giving what is called a beta-binomial distribution.

Let us consider data on the number of faults in rolls of fabric, as related to the lengths of the rolls given in Table 10.3. Here, we can use a simple linear Poisson regression for the counts, regressing them on the raw lengths. The data and the fitted curve are plotted in Figure 10.1. The regression model is estimated as

$$\log(\nu_j) = 0.9718 + 0.001\,93 x_j \tag{10.4}$$

and the deviance is 61.76 (65.76) with 30 d.f., indicating overdispersion. The difference in deviance from the null model with zero slope is 39.17

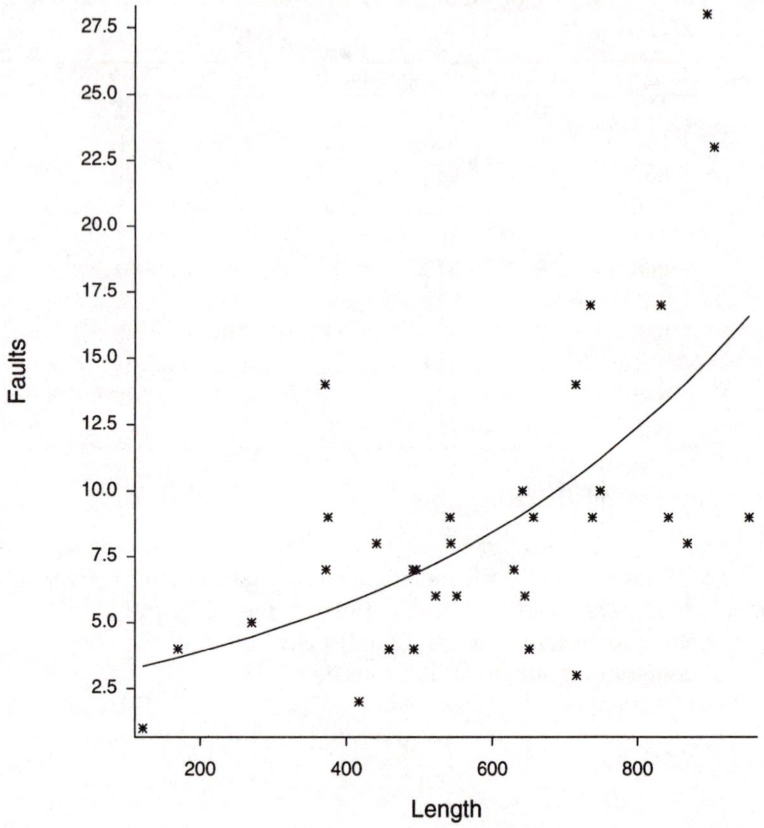

Fig. 10.1. Number of faults in rolls of fabric, from Table 10.3, with the simple linear Poisson regression curve (10.4).

with 1 d.f. From the plot, we can see that the variability in numbers of defaults seems to be increasing with the length of the roll, as would be expected for Poisson data. We also see that there is quite a bit of dispersion about the fitted line.

The heterogeneity factor for these data is $\hat{\phi} = 2.06$, which gives the scaled change for a model with zero slope as $39.17/2.06 = 19.01$, indicating that the latter model is implausible, even when overdispersion is taken into account.

The more elaborate approach is to use the beta-binomial distribution,

$$\Pr(y_{ij\bullet}) = \binom{n_{ij}}{y_{ij\bullet}} \int_{\Lambda} \frac{\lambda_i^{y_{ij\bullet}}(1-\lambda_i)^{n_{ij}-y_{ij\bullet}}\lambda_i^{\kappa_j-1}(1-\lambda_i)^{\upsilon_j-1}}{B(\kappa_j, \upsilon_j)} d\lambda_i$$

$$= \binom{n_{ij}}{y_{ij\bullet}} \frac{B(\kappa_j + y_{ij\bullet}, \upsilon_j + n_{ij} - y_{ij\bullet})}{B(\kappa_j, \upsilon_j)}$$

where $B(\cdot)$ is the beta function. We take a constant shape parameter, the intra-class correlation,

$$\rho = \frac{1}{\kappa + \upsilon + 1}$$

(GLIM4 macros for binomial and Poisson overdispersion are supplied in Appendix A.5.) Because one role of this parameter is to adjust for missing explanatory variables, only the estimate from the 'best' full model really measures intra-class correlation.

If we fit this model to the fabric data of Table 10.3, we obtain a deviance of 28.96 (34.96) with 29 d.f., the constant shape parameter, ρ, being estimated as 0.121. The difference in deviance from the null model with the same shape parameter is 19.31. The regression model is now estimated as

$$\log(\nu_j) = 1.003 + 0.001\,88x_j$$

little different from the previous one.

When applied to the egg hatching data of Table 10.1, for the factor level model, intra-class correlation is estimated as 0.0495 with a deviance of 56.53 (94.53) and 57 d.f. We can now fit all simpler models using this estimate obtained from the best model. Then, the deviance for the quadratic response surface model is 114.00 (136.0), for a difference of 57.47 with 8 d.f. If we look at the parameter estimates, in Table 10.2, we see that they have not changed very much from those for the binomial model. The standard errors for the beta-binomial model are slightly smaller than those which we obtained above using a heterogeneity factor. For the non-linear response surface model, the deviance is 67.24 (95.24), for a difference of 10.71 from the full factor model, with 5 d.f. We also see that the salinity·temperature interaction might be removed, with a further change in deviance of 3.71 (AIC 96.95).

A second common choice for $p(\cdot)$ is a normal distribution, usually for the canonical parameter of the distribution, $\log(\nu)$ for the Poisson and $\log[\pi/(1-\pi)]$ for the binomial distribution. Here, analytical results are not usually available and the integration has to be performed numerically. (A GLIM4 macro is given in Appendix A.6.)

Thus, returning to the fabric data, next we can fit the compound normal-Poisson model, which gives a deviance of 51.06 (57.06), which is not comparable, in value, to those given above, with 29 d.f. and a variance of 0.0984. The difference in deviance from the null model is 13.13 and the regression model is estimated as

$$\log(\nu_j) = 1.031 + 0.001\,73x_j$$

As would be expected, all three models give similar parameter estimates for the linear regression. The change in deviance for the effect of length is much less when overdispersion is taken into account, being least with the normal-Poisson model.

10.3 Clustering

In the counting process and Markov chain models of Chapters 7–9, time distinguished the different events on a unit. Occasionally, such events are distinct without being separated in time. Such nested or clustered study designs involve more than one count or more than one distinguishable event on each unit. Usually, some such response is observed on several subunits. If no explanatory variables distinguish the subunits, we have a simple clustered situation, while, if such variables are present, we have what is called a split plot. In such situations, the probability of events can differ among units, as differential frailty or proneness. At the same time, all events on a unit will be similarly inter-related.

One interesting result of the random effects approach, of giving a parameter of the regression model a random distribution, as in Equation (10.3), is that this model can be immediately extended to clustered data and to data with changing explanatory variables on each unit. Here, we have more than one count or distinguishable event for each unit so that the compound distribution of Equation (10.3) is multivariate. Again, two of the possibilities are conjugate and normal distributions. Because the integral contains a product of probabilities, care must be taken in models requiring numerical integration to avoid underflow. (The same GLIM4 macro, of Appendix A.6, as in the previous section, can be used for the normal compounding distribution.)

If the design for the explanatory variables is balanced, nested and split plot data are usually tabulated in a multi-dimensional contingency table, where one or more dimensions will represent repeated responses on the same unit.

We shall look at an example for binomial counts which can be displayed in a three-way contingency table. We have an overview of 25 clinical trials of a histamine H_2 antagonist compared with placebo in the treatment of upper gastrointestinal bleeding, as shown in Table 10.4. The study of such data is known as meta-analysis. Estimation of a common odds ratio in a set of 2×2 tables, such as this, is known as the Mantel–Haenszel problem.

The possible events are the responses, bleeding or not. Interest centres on differences between the drug and the control, while the differences among trials is a nuisance factor. Given the great variability in response among the trials, there is reason to suspect that the responses are not independent

Table 10.4. Overview of 25 trials of a histamine H_2 antagonist compared with placebo in treatment of upper gastrointestinal bleeding. (van Houwelingen and Zwinderman, 1993)

Treatment		Placebo		Year of
Bleeding	Total	Bleeding	Total	Publication
3	10	1	11	?
11	56	12	53	1979
8	50	16	50	1979
12	33	10	36	1979
2	14	6	15	1979
5	18	1	12	1979
9	40	11	48	1980
5	46	12	47	1980
10	21	8	19	1980
7	20	8	20	1980
9	45	3	43	1980
0	18	3	19	1980
11	36	5	26	1981
1	10	4	9	1981
0	14	0	14	1981
14	78	21	80	1982
3	34	13	31	1982
4	24	5	24	1982
50	259	51	260	1983
44	153	30	132	1984
16	106	15	107	1984
16	51	15	54	1984
6	31	12	29	1984
6	33	7	39	1984
2	15	3	14	1984

within a given trial.

A first reaction to such a table might be to analyse it in the standard way with a logistic or log linear model, looking at the differences in response between control and drug treatments. In such an approach, the variability among trials could be taken into account by the appropriate factor variable. Let us first try this. The deviance for a no-interaction logistic model with treatment and trial as explanatory variables

$$\texttt{TREAT} + \texttt{TRIAL}$$

is 41.19 (93.19) with 24 d.f., indicating a poor fit; the response to treatment seems to vary among trials. This deviance rises to 42.69 (92.69) when difference in treatment is removed from the model, for a difference of only 1.51 with 1 d.f., indicating that, on average, the effect of the drug is not very different from the control. The estimated effect is 0.123 (s.e. 0.100) on the logit scale. We might also add that the model with treatment differences, but without differences in trials, has a deviance of 104.86 (108.86) with 48 d.f. This also strongly indicates that the results from the 25 trials cannot be directly aggregated; we must keep the differences in trials in our model to control for the variability among them.

In this first model, we have taken into account the exact differences among the trials, but at the expense of a relatively large number of parameters: 24 for the differences among the trials. One important problem with this model is that the number of parameters is not fixed; it will increase with the number of observations, i.e. with each additional trial included. (This makes certain asymptotic inference procedures misleading.)

The idea of the random effects model is to allow for this heterogeneity among trials by the compounding with some distribution, in our case the normal. Then, the one parameter for the variance should have somewhat the same effect as the 24 trial contrasts above, giving the desired variability. The compound normal-binomial model gives an estimated treatment effect of 0.142 (s.e. 0.111). The standard deviation of the compounding normal distribution is estimated as 0.351. The change in deviance when the treatment effect is eliminated is 2.8 with 1 d.f., confirming the results above. Note that this model has 47 d.f. as compared to 24 for the standard logistic model. When a large number of units are involved, the simplification can be considerable as compared to a model where all differences among units are fitted.

However, in our particular example, the use of a random effects model is suspect because we saw, from the fixed effects model, that the treatment was not having the same effect on all trials. Thus, care must be taken in applying such models.

10.4 Symmetry models

A number of the models for square tables in Chapter 9 can be applied to clustered data, especially if there are few observations in each cluster. The most common occurrence is paired data, where the same response is observed on two objects grouped together, such as the two eyes of a person or twins in a family. Then, various symmetry models can be applied to try to account for the similarity within the cluster.

Let us look at data on vegetation sampling, where points are selected at random in a vegetation habitat. The vegetation at the point, and the closest neighbour of different species, are recorded. An example of such data are shown in Table 10.5. Inspecting the table, we see that *Dactylis glomerata* is most common, and that it very frequently has one of the

Table 10.5. Counts of nearest neighbours from contact sampling for vegetation species. (de Jong and Grieg, 1985)

	Neighbour									
Point	1	2	3	4	5	6	7	8	9	10
1 *Agropyron repens*	—	4	43	0	8	1	23	24	2	12
2 *Agrostis alba*	1	—	6	3	3	0	11	7	1	3
3 *Dact. glomerata*	41	10	—	10	106	12	123	156	13	60
4 *Holcus lanatus*	0	1	9	—	2	3	10	8	0	2
5 *Lolium perenne*	10	1	83	2	—	4	22	32	6	7
6 *Phleum pratense*	0	0	18	0	2	—	9	12	1	2
7 *Poa compressa*	15	5	59	4	15	3	—	52	4	29
8 *Trifolium repens*	10	11	55	4	12	3	34	—	3	11
9 *Plant. lancelotata*	1	0	7	1	2	2	4	2	—	1
10 *Tar. officinale*	9	0	32	0	4	0	11	12	2	—

other plants as neighbour. On the other hand, it is somewhat less often a neighbour of other plants.

The quasi-independence model

$$\texttt{POINT} + \texttt{NEIGHBOUR}$$

has a deviance of 4430.6 (4468.8) with 71 d.f., indicating a very poor fit. When a symmetry variable is added,

$$\texttt{POINT} + \texttt{NEIGHBOUR} + \texttt{SYM}$$

the deviance is 1803.8 (1911.8) with 36 d.f., still a very poor fit. Because the main diagonal is missing, this is just the quasi-symmetry model of Sections 5.3 and 9.2. Although this explains a substantial portion of the dependence structure, we can conclude that there is not a reciprocal relationship between randomly chosen plants and their nearest neighbours.

In other situations, a symmetry model, i.e. without the main effects, will be appropriate. An important special case of this, when there are only two categories, is known as the McNemar model.

10.5 Rasch model

In educational testing, items in a test often have a binary, true/false, response. Each subject replies to a series of questions making up the test. Responses will vary according to the ability of the subject and to the difficulty of each item. The latter are assumed to have some latent, or unobserved, variable in common. Rasch (1960) introduced a binary data model whereby the probability of response y_{ik} of the subject i to item k is given

Table 10.6. Responses to three questions on abortion in surveys conducted over three years. (Haberman, 1979, p. 482)

	Year		
Response	1972	1973	1974
YYY	334	428	413
YYN	34	29	29
YNY	12	13	16
YNN	15	17	18
NYY	53	42	60
NYN	63	53	57
NNY	43	31	37
NNN	501	453	430

by

$$\Pr(y_{ik}|\kappa_i) = \frac{e^{y_{ik}(\kappa_i - \upsilon_k)}}{1 + e^{\kappa_i - \upsilon_k}} \tag{10.5}$$

The data are represented by an $N \times R$ matrix of zeros and ones.

As with the clinical trials example in Section 10.3, the problem is that the number of subjects is not fixed. Thus, a standard log linear analysis of the frequencies of different combinations of responses is not acceptable. Rasch proposed to use a conditional likelihood approach, because conditioning on the marginal totals, $y_{i\bullet}$, eliminates the nuisance parameter, κ_i, from the likelihood function.

Subsequently, Tjur (1982) showed that the conditional model can be fitted as a log linear model. The margins for each item, $\mathtt{R}_k$, are fitted, as well as a factor variable for TOTAL score, with $R+1$ possible values,

$$\mathtt{R}_1 + \cdots + \mathtt{R}_R + \mathtt{TOTAL}$$

This can also be thought of as a model for quasi-independence in a $2^R \times (R+1)$ table containing structural zeros, because each combination of item responses only gives one score. It is also a generalization of the quasi-symmetry model, because all units with the same total correct response are treated symmetrically. Differences among groups can also be introduced into the model.

Let us look at the results of a survey about opinions on abortion which was carried out in three consecutive years, as shown in Table 10.6. The three questions concerned whether a pregnant woman should be able to obtain a legal abortion: 1. if she is married and does not want more children; 2. if the family has very low income and cannot afford any more children; and 3. if she is not married and does not want to marry the man. We are interested to see if one question has a positive response more frequently, if

Table 10.7. Parameter estimates from the Rasch model of Equation (10.6) applied to the abortion data of Table 10.6.

	Estimate	s.e.
1	6.217	0.045
YEAR(2)	−0.101	0.065
YEAR(3)	−0.153	0.066
R1	−0.756	0.066
R2	0.458	0.065
R3	−0.108	0.063
TOTAL(2)	−2.502	0.099
TOTAL(3)	−2.571	0.110
TOTAL(4)	0.000	—
TOTAL(2)·YEAR(2)	−0.080	0.150
TOTAL(3)·YEAR(2)	−0.064	0.162
TOTAL(4)·YEAR(2)	0.349	0.098
TOTAL(2)·YEAR(3)	0.076	0.147
TOTAL(3)·YEAR(3)	0.212	0.155
TOTAL(4)·YEAR(3)	0.365	0.099

one latent variable can describe attitudes to abortion, and if either evolves over the three years. This is not a panel study; different people were interviewed each year.

The simplest model is that the answers are the same for all three years:

$$\texttt{R1} + \texttt{R2} + \texttt{R3} + \texttt{TOTAL} + \texttt{YEAR}$$

where R1, R2, and R3 are binary indicators with one for yes, while TOTAL is the number of yeses given by a person. This has a deviance of 29.55 (45.55) with 16 d.f. If we add the dependence of the response on year,

$$(\texttt{R1} + \texttt{R2} + \texttt{R3} + \texttt{TOTAL}) * \texttt{YEAR}$$

the deviance is reduced to 4.85 (40.85) with 6 d.f., which is a very (too) acceptable model. Next we may ask if the distribution of the latent variable is the same in the different years. This is equivalent to checking if the total number of yeses varies over years, obtained by eliminating the interaction between YEAR and TOTAL. The model is now

$$(\texttt{R1} + \texttt{R2} + \texttt{R3}) * \texttt{YEAR} + \texttt{TOTAL}$$

with a deviance of 8.57 (36.57) and 10 d.f., a preferable model. The latent variable does not vary with year. We can also verify if the item characteristic curve is the same for all years by removing the interaction between the item responses and years

$$\texttt{R1} + \texttt{R2} + \texttt{R3} + \texttt{TOTAL} * \texttt{YEAR} \tag{10.6}$$

We obtain a deviance of 6.32 (34.32) with 10 d.f., showing an even more acceptable model. The questions appear to have the same relative frequency of positive response for all years. Finally, we check if all items have the same probability of positive response,

$$\texttt{TOTAL} * \texttt{YEAR}$$

which has a deviance of 145.93 (169.93) with 12 d.f., showing that they do not. The parameter estimates for our acceptable final model of Equation (10.6) are given in Table 10.7. Question two receives a positive response most often. As can also be seen from Table 10.6, zero or three yeses are more common than one or two. The number of people with three yeses (`TOTAL(4)`) is higher in 1973 and 1974 as compared to 1972, while the number with two is higher in 1974 than the previous years.

The plot of Cook's distances, in Figure 10.2, shows no problems. On the other hand, the normal probability plot, in Figure 10.3, is not very straight, perhaps due to overfitting.

As we can see, this model has much wider application than simply to the analysis of the results of individual questions on a test. We have a multivariate binary response for a series of units, under a number of conditions, so that the responses on a unit are completely differentiated. The indexing of the units is a nuisance parameter, in which we are not interested, and the binary responses are not ordered (in time). In such situations, the model originally proposed by Rasch (1960) is especially interesting because it can easily be fitted by standard log linear model techniques. The model can also be extended to responses with several categories (Conaway, 1989). Notice also that, if instead the questions are ordered by difficulty, the Guttman scale model of Section 5.5 can be applied to such data.

10.6 Bradley–Terry model

Occasionally, people may be asked to make a series of comparisons between pairs of objects, stating which is preferred. The result will be a square table showing how many individuals prefer each object as opposed to each other. The two variables are 'prefer' and 'not prefer', each with as many categories as there are objects to compare. The idea is to rank the objects in order of global preference for the group of people. If all people rank all objects, the rank is simply obtained from the number of positive preferences expressed (as in the ranking of teams in some sport). With unequal numbers, the problem is more complex.

We are concerned with ranking preferences, so that ties may be ignored because they provide no information about ordering among the objects. However, if there is a relatively large number of ties, the validity of the

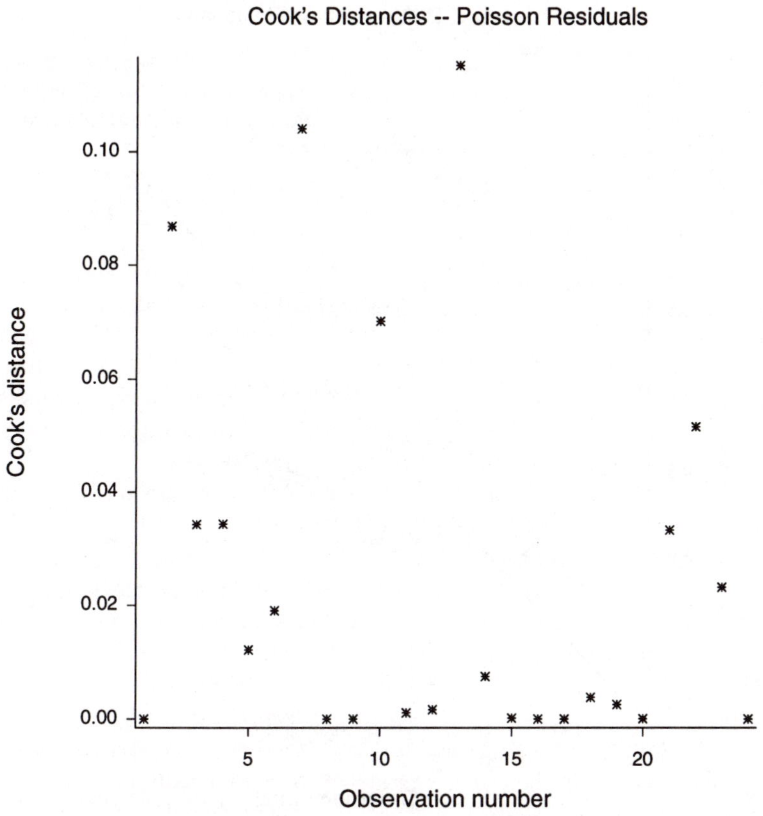

Fig. 10.2. Cook's distances from the model of Equation (10.6) for the abortion data of Table 10.6.

study will be questionable. Although we now have an incomplete table, with the diagonal missing, our problem is not resolved. We may construct a new table with one dimension being the object preferred and the other being the pair compared. For example, with four objects, we have

	1	2	3	4
(1,2)	n_{12}	n_{21}	—	—
(1,3)	n_{13}	—	n_{31}	—
(1,4)	n_{14}	—	—	n_{41}
(2,3)	—	n_{23}	n_{32}	—
(2,4)	—	n_{24}	—	n_{42}
(3,4)	—	—	n_{34}	n_{43}

This is now a different incomplete table, to which we could fit a model of quasi-independence. However, notice that one margin, for object preferred,

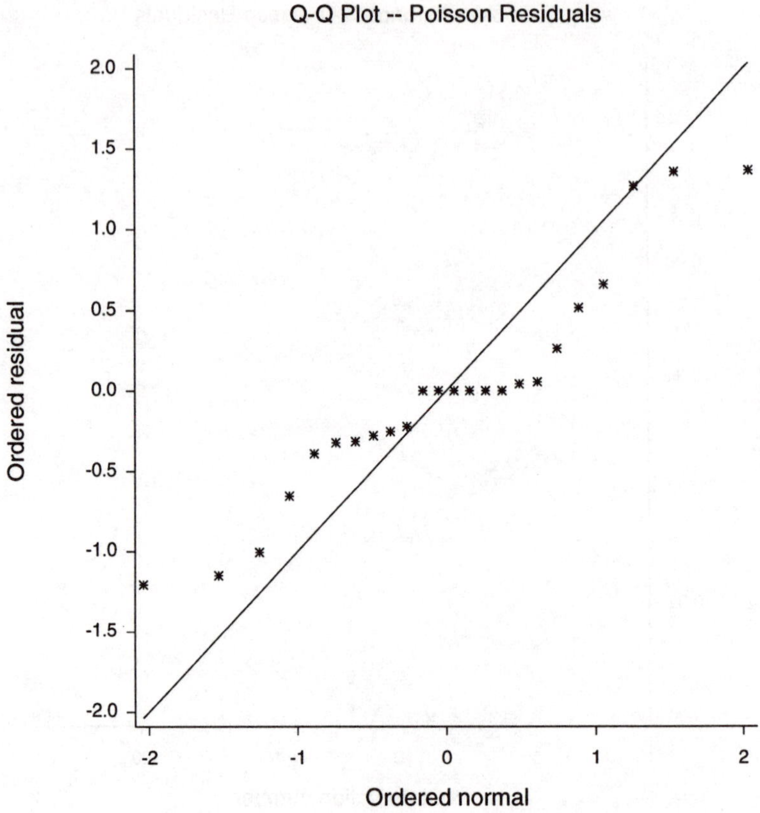

Fig. 10.3. Normal probability plot from the model of Equation (10.6) for the abortion data of Table 10.6.

is identical to that in the original table. The second margin corresponds to a symmetry variable for the original table. We are checking to ensure that the general ranking of all objects does not depend on the specific pairs of comparisons made by the individual. If this model is acceptable, the parameters for the variable, object preferred (the columns of the table), give the rank.

This construction gives us a hint that the same model may also be developed in another way which does not require the table to be reconstructed. Instead, aspects of its symmetry are used. The factor variable (SYM) required for this model, corresponding to the one margin of the new table, is just that used in the symmetry models of Sections 5.3, 9.2, and 10.4, with zero weight for the diagonal. However, this Bradley–Terry model contains one set of mean parameters, that for preferences, whereas those

Table 10.8. Preferences for brands of carbon paper by typists. (Glenn and David, 1960, from Freund and Jackson)

	Not preferred				
Preferred	1	2	3	4	5
1	—	17	3	24	13
2	8	—	8	17	8
3	21	18	—	26	16
4	4	7	1	—	4
5	13	16	10	23	—

in the previous sections had either none or two. The model to be fitted is, then,

$$\texttt{PREF} + \texttt{SYM} \tag{10.7}$$

where PREF is a factor variable indicating the object preferred. Thus, we are modelling quasi-independence in the reconstructed table, but without actually setting it up.

Data are available on comparisons between five brands of carbon paper as judged by 30 typists and shown in Table 10.8. We shall apply the model of Equation (10.7) to this table. With a deviance of 7.39 (35.39) and 10 d.f., the model fits very well. From the parameter estimates for PREF, (0.000, −0.466, 0.957, −1.620, 0.215), the third brand is found to be preferred, with the fourth being last. We can check for equality of preferences by removing PREF from the model. The change in deviance, 82.97 (AIC 110.36) with 4 d.f., shows that such a model is not acceptable.

For the preference model of Equation (10.7), Cook's distances, in Figure 10.4, indicate that the preferences of paper 3 over 1 and 2 fit badly to the model. On the other hand, the normal probability plot (not shown) is reasonably straight, indicating a good model.

10.7 Markov chains with clustering

The example above for the Rasch model involved observations over time, but on different individuals. If several distinct binary responses are recorded at each point in time, Markov chain models can be combined with the Rasch model to take into account the serial dependence and the clustering simultaneously. Because the techniques are similar to those which we have already seen for these two models, we can immediately look at an example.

Let us consider a study of the infection of the teats of dairy cattle in Denmark. Milk samples were taken every three months, from 5 to 13 times, from 1233 cattle in 67 herds, and the presence or absence of bacteria recorded for each teat, as given in Table 10.9. We shall assume a first-order stationary Markov chain model and we shall ignore any effect of clustering

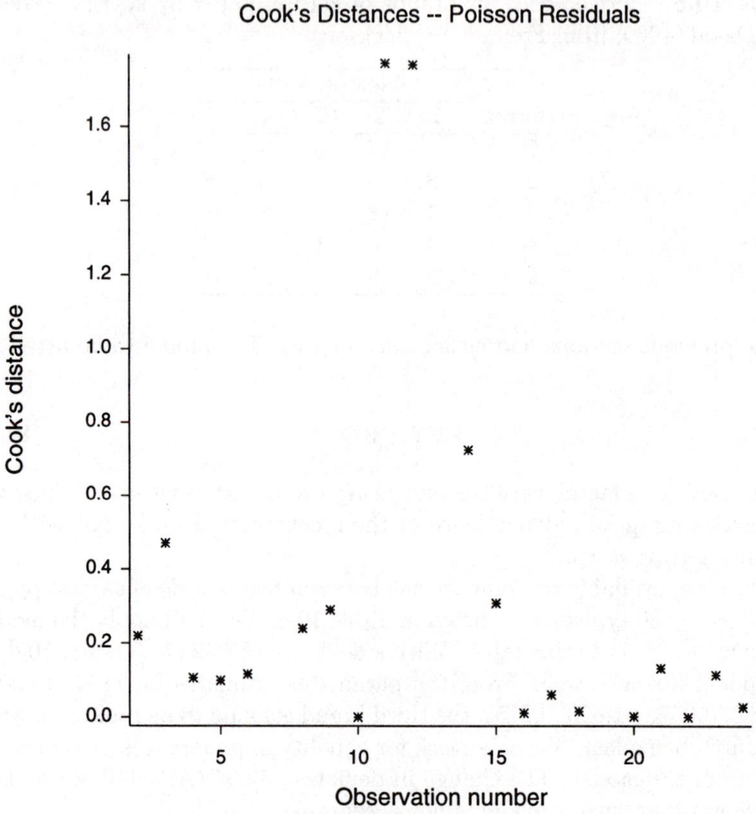

Fig. 10.4. Cook's distances from the Bradley–Terry model of Equation (10.7) for the typist data of Table 10.8.

at the herd level for which no information is available.

We can start with the model of independence, where present infections do not depend on the previous state of each teat:

$$\mathtt{T1RF} + \mathtt{T1LF} + \mathtt{T1RR} + \mathtt{T1LR} + \mathtt{T1TOTAL} + \mathtt{T0}$$

Here, T0 is a 16 level factor variable for all possible combinations of the previous state, T1TOTAL is the number of teats presently infected, and T1RF, etc., are 1/0 variables indicating whether or not each teat is currently infected. This model has a deviance of 5234.3 (5280.3) with 233 d.f.

The model for the complete first-order Markov chain, with clustering,

$$(\mathtt{T1RF} + \mathtt{T1LF} + \mathtt{T1RR} + \mathtt{T1LR} + \mathtt{T1TOTAL}) * \mathtt{T0}$$

has a deviance of 220.97 (476.97) with 128 d.f. (The saturated model has an AIC of 512.) If the latent variable does not vary with the previous state of each teat, we have

$$(\texttt{T1RF} + \texttt{T1LF} + \texttt{T1RR} + \texttt{T1LR}) * \texttt{T0} + \texttt{T1TOTAL}$$

which has a deviance of 329.67 (495.67) with 173 d.f., so that there is dependence. If we have clustering, but the present state does not depend on the previous one,

$$\texttt{T1RF} + \texttt{T1LF} + \texttt{T1RR} + \texttt{T1LR} + \texttt{T1TOTAL} * \texttt{T0}$$

the deviance is 2679.7 (2845.7) with 173 d.f., showing that the present state of each teat depends on the previous state of the teats. However, it may only depend on its own previous state, and not that of the other teats,

$$\texttt{T0RF} * \texttt{T1RF} + \texttt{T0LF} * \texttt{T1LF} + \texttt{T0RR} * \texttt{T1RR} + \texttt{T0LR} * \texttt{T1LR} + \texttt{T1TOTAL} * \texttt{T0} \qquad (10.8)$$

This model has a deviance of 278.87 (452.87) with 169 d.f., showing that dependence on other teats is not necessary. Another possibility is that the present state of each teat only depends on the total number of infected teats at the previous time,

$$(\texttt{T1RF} + \texttt{T1LF} + \texttt{T1RR} + \texttt{T1LR}) * \texttt{T0TOTAL} + \texttt{T1TOTAL} * \texttt{T0}$$

but this has a deviance of 2664.9 (2854.9) with 161 d.f. Finally, we check if all teats have the same probability of infection,

$$\texttt{T1TOTAL} * \texttt{T0}$$

which has a deviance of 2826.2 (2986.2) with 176 d.f.

We now return to our model in Equation (10.8). If we look at the parameter values (not shown), we see that the parameters for the transition probabilities (the four interactions terms between T0ab and T1ab for the four teats) have similar estimates, as do those for T1RF,T1LF,T1RR,T1LR. We may set the transition probabilities to be equal by constructing a new variable which is the sum of the products of the states of each teat at the two times:

$$\texttt{T1RF} + \texttt{T1LF} + \texttt{T1RR} + \texttt{T1LR} + \texttt{SUMTOT1} + \texttt{T1TOTAL} * \texttt{T0}$$

This model has a deviance of 286.79 (454.79) with 172 d.f., indicating that all transitions are not identical. If we separate front and rear teats,

Table 10.9. Changes during three-month periods in presence of bacteria in milk samples from cow's right front, left front, right rear, and left rear teats. (Gottschau, 1994)

	Present state							
Previous	0000	0001	0010	0011	0100	0101	0110	0111
0000	3253	289	284	79	193	37	34	24
0001	213	281	35	33	15	30	4	3
0010	193	28	229	46	20	4	27	12
0011	66	33	27	83	3	7	3	10
0100	144	22	18	5	169	30	14	9
0101	27	25	8	5	20	42	1	10
0110	19	2	22	2	18	2	34	9
0111	9	5	8	11	8	8	7	17
1000	142	12	10	10	14	4	3	4
1001	22	17	3	5	1	2	0	0
1010	23	5	23	11	1	1	0	1
1011	12	5	6	7	3	3	3	4
1100	34	5	3	3	15	5	6	2
1101	11	7	0	0	2	6	0	0
1110	15	2	7	3	4	0	4	0
1111	18	3	7	11	4	4	6	12
	1000	1001	1010	1011	1100	1101	1110	1111
0000	174	43	45	17	44	15	19	24
0001	19	20	6	10	5	11	3	6
0010	9	4	29	12	4	1	8	12
0011	9	3	5	12	0	4	5	12
0100	14	1	2	0	25	5	6	5
0101	5	3	1	0	4	10	0	4
0110	1	1	1	1	1	1	5	7
0111	0	1	1	2	2	3	4	12
1000	161	15	24	8	21	3	9	7
1001	15	26	3	10	3	4	2	4
1010	13	3	33	10	6	6	9	8
1011	8	5	5	19	2	1	5	7
1100	22	7	4	0	32	12	6	9
1101	4	5	0	1	6	10	4	5
1110	6	0	12	3	8	1	18	8
1111	7	2	5	7	7	7	4	28

$$\text{T1RF} + \text{T1LF} + \text{T1RR} + \text{T1LR} + \text{SUMTOT1F} + \text{SUMTOT1R} + \text{T1TOTAL} * \text{TO}$$

we obtain a deviance of 280.29 (450.29) with 171 d.f. We may also try equating the probabilities of infection in the four teats by creating a variable which is the sum of `T1RF,T1LF,T1RR,T1LR`:

$$\text{SUMT1} + \text{SUMTOT1F} + \text{SUMTOT1R} + \text{T1TOTAL} * \text{TO}$$

Here, we find a deviance of 354.05 (518.05) with 174 d.f. However, if we also distinguish between front and rear in the probability of infection,

$$\text{SUMT1F} + \text{SUMT1B} + \text{SUMTOT1F} + \text{SUMTOT1R} + \text{T1TOTAL} * \text{TO}$$

we obtain a deviance of 280.68 (446.68) with 173 d.f. Thus, the log odds of risk of infection is different between front (1.047) and rear (1.408) teats, being higher at the rear. In the same way, the dependence of this risk on infection three months before also differs (front: 2.293; rear: 2.071).

Thus, we see how the Rasch model provides a simple way of handling clustering in categorical data, even when dependence over time is present.

10.8 Exercises

(1) Batches of *Orobanche cernua* seeds were placed in bean root extract at three dilutions and the number of germinated seeds counted (Crowder, 1978):

Dilution					
1/1		1/25		1/625	
Germinate	Total	Germinate	Total	Germinate	Total
2	43	17	19	11	13
9	51	43	56	47	62
5	44	79	87	90	104
16	71	50	55	46	51
2	24	9	10	9	11
0	7				

(a) Develop a suitable model which takes into account differences among batches.

(b) Does the dilution level have an effect on germination?

(2) The table below shows the effects of salinity and temperature on the proportion of eggs of English sole hatching in four independent tanks under each combination of conditions (Lindsey *et al.*, 1970; Lindsey 1975).

Salinity	Temperature	Hatch	Total	Hatch	Total
15	4	236	666	203	724
		183	764	212	723
15	8	600	656	697	747
		615	746	641	703
15	12	407	566	343	603
		365	560	302	394
25	4	203	717	177	782
		155	852	138	590
25	8	591	621	564	640
		714	754	532	570
25	12	475	622	465	645
		506	608	415	532
35	4	1	738	3	655
		10	742	3	743
35	8	526	616	419	467
		410	484	374	606
35	12	272	362	352	478
		392	590	382	459
10	10	303	681	329	710
		262	611	301	700
10	6	277	757	234	681
		263	647	287	801
40	10	387	450	389	553
		388	564	318	604
40	6	276	662	247	542
		248	527	149	591
20	10	351	391	559	650
		527	603	476	548
20	6	585	643	620	671
		437	497	667	771
30	10	447	491	462	530
		475	545	499	556
30	10	522	573	615	680
		539	581	517	561
30	6	563	666	600	704
		562	656	615	723

Find an appropriate model to describe these data.

(3) The effect of rocking on the crying of very young babies was studied in a hospital nursery (Cox, 1966). The numbers of babies crying under control and experimental conditions were recorded over 18 days, as in the following table:

	Control		Experimental	
Day	Crying	Total	Crying	Total
1	3	8	1	1
2	2	6	1	1
3	1	5	1	1
4	1	6	0	1
5	4	5	1	1
6	4	9	1	1
7	5	8	1	1
8	4	8	1	1
9	3	5	1	1
10	8	9	0	1
11	5	6	1	1
12	8	9	1	1
13	5	8	1	1
14	4	5	1	1
15	4	6	1	1
16	7	8	1	1
17	4	6	0	1
18	5	8	1	1

(a) Is there any evidence of a difference when rocking is used?

(b) Is there any variation over days?

(4) The following table shows the results of a multicentre randomized clinical trial to study the efficacy of two topical cream preparations in curing infection. The unit of observation is the clinic and the two treatments are a cream for curing infection and the control (Beitler and Landis, 1985).

Cream		Control	
Favourable	Unfavourable	Favourable	Unfavourable
11	25	10	27
16	4	22	10
14	5	7	12
2	14	1	16
6	11	0	12
1	10	0	10
1	4	1	8
4	2	6	1

(a) Determine which treatment works better.

(b) Is there any difference in results among clinics?

(5) Results of eight case-control studies on the association between lung cancer and smoking are presented in the table below (Christensen, 1989, p. 193, from Dorn).

Control		Case	
Non-smoker	Smoker	Non-smoker	Smoker
14	72	3	83
43	227	3	90
19	81	7	129
81	534	18	459
61	1296	7	1350
27	106	3	60
56	462	19	499
28	259	5	260

(a) Is there any evidence of a link between cancer and smoking?
(b) Is there any difference in results among the studies?

(6) In a study of pregnancy and child development in Germany, perinatal mortality and father's smoking habits (number of cigarettes per day) were recorded in 9 clinics (Wermuth, 1976):

Smoking	Deaths	Total	Deaths	Total	Deaths	Total
	Clinic 1		Clinic 2		Clinic 3	
None	5	160	4	109	9	405
1–10	0	51	1	63	3	164
> 10	5	44	3	95	6	270
	Clinic 4		Clinic 5		Clinic 6	
None	7	188	4	160	5	129
1–10	0	77	2	97	0	56
> 10	10	151	4	104	6	105
	Clinic 7		Clinic 8		Clinic 9	
None	2	176	16	538	8	120
1–10	2	60	7	231	4	62
> 10	1	114	10	222	6	94

(a) How does perinatal death depend on the father's smoking habits?
(b) Does it vary among the clinics?

(7) The following table gives the classification of hospitalized sibling pairs, the elder sibling first, by schizophrenic diagnosis and by sex (Cohen, 1976).

	Diagnosis			
Sex	00	01	10	11
MM	2	1	1	13
MF	4	1	3	5
FM	3	3	1	4
FF	15	8	1	6

0: not schizophrenic
1: schizophrenic

Find an appropriate model to describe these data.

(8) The table below gives the diagnostic classification regarding multiple sclerosis performed by two neurologists for two populations (Landis and Koch, 1977).

	Winnipeg neurologist's diagnostic				
	Class	1	2	3	4
	Winnipeg patients				
	1	38	5	0	1
	2	33	11	3	0
	3	10	14	5	6
New Orleans neurologist's diagnostic	4	3	7	3	10
	New Orleans patients				
	1	5	3	0	0
	2	3	11	4	0
	3	2	13	3	4
	4	1	2	4	14

Find an appropriate model to describe these data.

(9) The following table gives the results of eye tests of employees aged 30–39 in Royal Ordinance factories in Britain between 1943 and 1946 (Stuart, 1955).

Right eye grade	Left eye grade				
	4	3	2	1	Total
	Female				
4	1520	266	124	66	1976
3	234	1512	432	78	2256
2	117	362	1772	205	2456
1	36	82	179	492	789
Total	1907	2222	2507	841	7477
	Male				
4	821	112	85	35	1053
3	116	494	145	27	782
2	72	151	583	87	893
1	43	34	106	331	514
Total	1052	791	919	480	3242

(a) Find an appropriate model to describe these data.
(b) Are there differences between men and women?

(10) The Danish Welfare Study looked at various personal hazards (Andersen, 1991, p. 483). The question on work asked if the person was often exposed to noise, bad light, toxic substances, heat, and dust. That for psychic inconvenience asked if the person suffered from neuroses, sensitivity to noise, sensitivity to minor difficulties, troubling

thoughts, and shyness, while that for physical inconveniences asked about diarrhea, back pain, colds, coughing, and headaches.

Answer	Work	Psychic	Physical
YYYYY	70	34	16
YYYYN	15	10	5
YYYNY	34	21	24
YYYNN	6	17	16
YYNYY	39	17	11
YYNYN	21	4	12
YYNNY	38	15	79
YYNNN	49	20	98
YNYYY	103	63	18
YNYYN	39	21	11
YNYNY	129	45	19
YNYNN	66	38	15
YNNYY	115	42	9
YNNYN	116	29	9
YNNNY	217	65	54
YNNNN	409	92	97
NYYYY	4	14	37
NYYYN	8	4	38
NYYNY	7	35	73
NYYNN	3	32	82
NYNYY	12	21	55
NYNYN	22	12	70
NYNNY	16	99	365
NYNNN	60	172	689
NNYYY	24	46	30
NNYYN	27	27	61
NNYNY	54	115	84
NNYNN	99	176	178
NNNYY	63	98	44
NNNYN	193	107	131
NNNNY	168	776	454
NNNNN	1499	2746	2267

(a) Fit the Rasch model to these data.
(b) Are the results the same for all three questions?
(c) Is the distribution of the latent variable the same for the different questions?
(d) Is the item characteristic curve the same for all questions?

(11) Data on responses (C: correct, W: wrong) to four questions from the arithmetic reasoning test on the Armed Services Vocational Apti-

tude Battery, with samples from white and black males and females (Mislevy, 1985) are reproduced below.

	White		Black	
Response	M	F	M	F
CCCC	86	42	2	4
WCCC	1	7	3	0
CWCC	19	6	1	2
WWCC	2	2	3	3
CCWC	11	15	9	5
WCWC	3	5	5	5
CWWC	6	8	10	10
WWWC	5	8	5	8
CCCW	23	20	10	8
WCCW	6	11	4	6
CWCW	7	9	8	11
WWCW	12	14	15	7
CCWW	21	18	7	19
WCWW	16	20	16	14
CWWW	22	23	15	14
WWWW	23	20	27	29

(a) Fit the Rasch model to these data.
(b) Are the test results the same for all four groups?
(c) Is the distribution of the latent variable the same in the different groups?
(d) Is the item characteristic curve the same for all groups?
(e) Do all items have the same difficulty?

12) Andersen (1980, p. 357) provides a table of preferences expressed for a series of six collective facilities in a Danish municipality. Unfortunately, he does not include information about which facilities were compared.

	Not preferred					
Preferred	1	2	3	4	5	6
1	—	29	25	22	17	9
2	49	—	35	34	16	14
3	50	42	—	40	22	15
4	54	43	37	—	33	16
5	61	61	54	44	—	27
6	69	64	63	62	51	—

Determine the order of preference for the facilities.

13) Six brands of chocolate milk pudding were compared in pairs (Davidson, 1970):

	Not preferred					
Preferred	1	2	3	4	5	6
1	—	19	16	18	13	18
2	22	—	19	23	16	22
3	19	23	—	19	16	13
4	23	19	20	—	17	14
5	19	20	15	14	—	11
6	21	20	18	19	21	—

Determine the order of preference for the puddings.

Appendix A
GLIM macros

A.1 Log multiplicative model

```
!  The macro L1OV calculates the appropriate scale for the
!relationship  between a (combination of) nominal and an ordinal
!variable in a log-linear model.
!  Set up the model as usual with $SLength, $Yvariate, $ERror P,
!$FActor. The two variables concerned must both be declared in
!$FActor. Type $Use L1OV with the two variable names (second one
!ordinal).
!  The scale is returned in a new quantitative variable, SCALE_,
!which may be used in subsequent $Fits, for example, if a combined
!nominal variable is fitted as separate variables.
!  Macros used: L1OV, PRC1, ITER
!
$Macro (Argument=ORD1,ORD2 Local=FACV_,SC1_,J_,SC2_,I_,LINV_) L1OV !
$CAlculate %Z8=ORD1(1)!                  calculate dimensions of table
: %Z8=%IF(ORD1>%Z8,ORD1,%Z)!
: %Z9=ORD2(1)!
: %Z9=%IF(ORD2>%Z9,ORD2,%Z9)!
: %Z4=%Z6=0!
: %Z2=10!
: LINV_=2*ORD2-%Z9-1!                        calculate linear variable
$PRint 'Independence Model' :!
$Fit ORD1+ORD2!
$Display E!
$PRint 'Linear Effects Model' :!
$Fit ORD1+ORD2+ORD1.LINV_!
$Display E!
$OUtput!                  stop output during iterative calculations
$TRanscript!
$Variate %Z8 SC1_ J_!
: %Z9 SC2_ I_!
$CAlculate FACV_=ORD1!
: J_=%GL(%Z8,1)+%Z8!
: I_=%GL(%Z9,1)+%Z9!
$Argument ITER ORD1 ORD2 FACV_ SC2_ I_ J_ SCALE_!
$WHile %Z2 ITER!
```

```
$Fit ORD1+ORD2+ORD1.SCALE_!                              final fit for scale
$CAlculate %Z4=%DF-%Z9+2!                                               d.f.
: %Z3=SC2_(1)!                   standardize scale to lie between 0 and 1
: %Z7=SC2_(%Z9)-%Z3!
: SC2_=(SC2_-%Z3)/%Z7!
$OUtput %POC!                                                 restart output
$TRanscript F H I O W!
$PRint 'Scale for ordinal variable'!                         print out model
: SC2_ :!
: 'Log Multiplicative Model' :!
: 'scaled deviance = ' *4 %DV ' at cycle' *-2 %Z6!
: '           d.f. = ' *-2 %Z4 :!
$SWitch %Z5 PRC1!
$Display E!
$DElete FACV_ SC1_ J_ SC2_ I_ LINV_!
$$Endmac!
!
$Macro (Argument=ORD1,ORD2,FACV_,SC2_,I_,J_,SCALE_) ITER !
$CAlculate %Z6=%Z6+1!                               iterative fitting macro
: %Z5=%DV!
$Fit FACV_+ORD2+FACV_.ORD2!                             estimate first scale
$EXTract %PE!
$CAlculate %Z3=%PE(2)!
: SC2_=(I_/=%Z9+1)*%PE(I_)+%Z3!
: SCALE_=(ORD2/=1)*%PE(ORD2+%Z9)+%Z3!
: %Z3=SC2_(1)!
: %Z7=SC2_(%Z9)-%Z3!
: SCALE_=(SCALE_-%Z3)/%Z7!
$Fit SCALE_+ORD1.SCALE_+ORD1!                          estimate second scale
$EXTract %PE!
$CAlculate %Z3=%PE(2)!
: SC1_=(J_/=%Z8+1)*%PE(J_)+%Z3!
: FACV_=(ORD1/=1)*%PE(ORD1+%Z8)+%Z3!
: %Z3=SC1_(1)!
: %Z7=SC1_(%Z8)-%Z3!
: FACV_=(FACV_-%Z3)/%Z7!
: %Z5=(%Z5-%DV)/%DV!                                    test for convergence
: %Z2=%IF(%Z5=(%Z5*%Z5>.00001),%Z2-1,0)!
$$Endmac!
!
$Macro PRC1 !                          message to print if no convergence
$PRint '     (no convergence yet)' :!
$$Endmac!
```

A.2 Proportional odds models

```
!  The macro POOV fits a proportional odds model for an ordinal
!dependent variable with grouped frequency data.
```

```
!  $CAlculate %N=number of independent !variables, %K=number of
!categories of the dependent variable (%N+%K <= 10), and %L=number
!of lines in the table (table size = %Lx%K).
!  Then, type $Use POOV with the names of the frequency vector and
!up to 7 independent variables.
!  The ordinal dependent variable must vary most quickly in the
!frequency vector. All independent variables must be continous or
!binary ($FActor cannot be used).
!  The first %K-1 parameter estimates refer to the odds for
!categories of the dependent variable, the last %N to the
!independent variable.
!  See Hutchison, D. (1985) "Ordinal variable regression using the
!McCullagh (proportional odds) model." GLIM Newsletter 9: 9-17.
!  Macros used: POOV, INDV, IND1, STEP, CM, ETML, WMU, PARA, FVDR,
!VADI, MPE, INIT

$Macro (Argument=FREQ
  Local=EXPV_,C1_,C2_,C3_,C4_,C5_,C6_,C7_,C8_,C9_) POOV
$DElete Y_ N_ I_ BL_ EXPV_ TOT_ J_ C1_ C2_ C3_ C4_ C5_ C6_ C7_ C8_!
  C9_ !
$CAlculate %Z7=9!
: %Z2=%K*%L!
$Argument STEP FREQ!
: INDV %2 %3 %4 %5 %6 %7 %8 %9!
: IND1 C1_ C2_ C3_ C4_ C5_ C6_ C7_ C8_ C9_!
$CAlculate I_=%GL(%L,%K)!       create indices to manipulate vectors
: %Z4=(%K-1)*%L!
$Variate %L TOT_!
: %SL J_!
$SLength %Z4!
$CAlculate TOT_=0!                                  calculate totals
: TOT_(I_)=TOT_(I_)+FREQ!
: EXPV_=%GL(%L,1)!
: %Z1=%K-1!                              calculate response variable
: Y_=0!
$WHile %Z1 STEP!
$DElete I_ J_!
$CAlculate %Z4=%SL*%Z7!   initialize to create explanatory variables
$Variate %Z4 BL_ N_ I_ J_!
$CAlculate I_=%GL(%Z7,1)!
: N_=(I_==%GL(%K-1,%Z7*%K))!
: J_=(%GL(%K,%Z7)-1)*%K+%GL(%K,%K*%Z7)!
: %Z8=%N!
$WHile %Z8 INDV!          set up explanatory variables in new vectors
$Variate %SL C1_ C2_ C3_ C4_ C5_ C6_ C7_ C8_ C9_!
$CAlculate %Z8=%Z7!
$WHile %Z8 IND1!
$DElete N_ I_ BL_ J_!
```

```
$Yvariate Y_!                                    set up approximate model
$ERror B N_!
$CAlculate N_=TOT_(EXPV_)!                       transform totals vector
: BL_=%GL(%K-1,%L)!
$PRint 'Proportional Odds Model' :!              print transformed table
: '      R         N         BL_       EXPV_'!
$Look (S=-1) Y_ N_ BL_ EXPV_!
$DElete BL_ EXPV_!
$OUtput!                   stop output during iterative calculations
$TRanscript!
$Fit C1_+C2_+C3_+C4_+C5_+C6_+C7_+C8_+C9_-1!      fit approximate model
$EXTract %PE!                                    for initial estimates
$CAlculate PE_=%PE!
$DElete Y_ N_ C1_ C2_ C3_ C4_ C5_ C6_ C7_ C8_ C9_!
$SLength %Z2!
$CAlculate %Z4=%Z2*%Z7!  initialize to create independent variables
$Variate %Z4 N_ I_ BL_ J_!
$CAlculate I_=%GL(%Z7,1)!
: N_=(I_==%GL(%K,%Z7))*(I_<%K)!
: J_=%GL(%SL,%Z7)!
: %Z8=%N!
$WHile %Z8 INDV!         set up independent variables in new vectors
$DElete J_!
$CAlculate Y_=TOT_(%GL(%L,%K))!
: J_=%GL(%SL,1)!
$DElete I_ BL_ TOT_!
$Yvariate FREQ!                                       set up exact model
$ERror P!
$LInk O FVDR!
$METhod * MPE!
$INitial INIT!
$SCale 1!
$Argument FVDR C1_ C2_ C3_ C4_ C5_ C6_ C7_ C8_ C9_!
: WMU %2 %3 %4 %5 %6 %7 %8 %9 %1!
$CAlculate %LP=C1_=C2_=C3_=C4_=C5_=C6_=C7_=C8_=C9_=0!
$OUtput %POC!                                             restart output
$TRanscript F H I O W!
$Fit C1_+C2_+C3_+C4_+C5_+C6_+C7_+C8_+C9_-1!              fit exact model
$Display E!
$$Endmac!
!
$Macro (Argument=FREQ) STEP !            fills new vector with values
$CAlculate J_=(%GL(%K,1)<=%K-%Z1)*(%GL(%L,%K)+(%K-1-%Z1)*%K)!
: Y_(J_)=Y_(J_)+FREQ!
: %Z1=%Z1-1!
$$Endmac!
!
```

```
$Macro INDV !         first step to create new independent variables
$CAlculate BL_=(I_==%K+%N-%Z8)*J_!
: N_=N_+%%Z8(BL_)!
: %Z8=%Z8-1!
$$Endmac!
!
$Macro IND1 !         second step to create new independent variables
$CAlculate BL_=(I_==%Z8)*%GL(%SL,%Z7)!
: %%Z8(BL_)=N_!
: %Z8=%Z8-1!
$$Endmac!
!
$Macro INIT !
$CAlculate %ETA=%LOG(%YV+0.5)!
$Endmac
!
$Macro MPE !
$EXTract %PE!
$CAlculate PE_=%PE!
$Endmac
!
$Macro FVDR !     calculate fitted values and deta by dgamma for own
$CAlculate %DR=1!
: %Z6=%Z7!
: %ETA=40*(%GL(%K,1)==%K)!
$WHile %Z6 ETML!
$CAlculate TOT_=%EXP(%ETA)/(1+%EXP(%ETA))!
: I_=TOT_/(1+%EXP(%ETA))!
: %Z5=1!
$Use CM TOT_ %FV!                         calculate fitted values vector
$CAlculate %Z5=2!
$Use CM %ETA %LP!                      calculate linear predictor vector
$CAlculate %Z6=%Z7!
$Use WMU!
$$Endmac!
!
$Macro ETML !                                             calculate eta
$CAlculate %ETA=%ETA+PE_(%Z6)*N_(%Z7*(J_-1)+%Z6)!
: %Z6=%Z6-1!
$$Endmac!
!
$Macro CM !                   set up fitted values and linear predictor
$CAlculate %1=((%Z5==1)+(%Z5==2)*I_)*%1!
: %2=(%1-%1(J_-1)*(%GL(%K,1)/=1))*Y_!
$$Endmac!
!
$Macro WMU !                           calculate new parameter estimates
```

```
$CAlculate BL_=N_((J_-1)*%Z7+(%Z7-%Z6+1))!
$Use CM BL_ %9!
$CAlculate %Z8=((%Z6=%Z6-1)>0)!
$SWitch %Z8 WMU!
$$Endmac!
```

A.3 Marginal homogeneity model

```
!  The macro MHCT fits a marginal homogeneity model to a square
!2-way table (max. 10x10).
!  Set up the log linear model as usual with $SLength, $Yvariate,
!$ERror P, $FActor. Type $Use MHCT with the 2 factor variables.
!  Macros used: MHCT, MARG, ITMH, PRCY
!
$Macro (Local=UN_ Argument=M1,M2) MHCT !
$OUtput!                       stop output during iterative calculations
$TRanscript!
$DElete UN_ FAC1_ FAC2_ RES_ PW_ FIT_!
$CAlculate %Z6=M1(%SL)!                                    size of table
: FAC1_=M1!                                           initialize vectors
: FAC2_=M2!
: FIT_=RES_=%YV!
: C1_=C2_=C3_=C4_=C5_=C6_=C7_=C8_=C9_=0!
: %Z1=10!                                           number of iterations
: %Z5=0!
$Argument ITMH C1_ C2_ C3_ C4_ C5_ C6_ C7_ C8_ C9_!
: MARG %1 %2 %3 %4 %5 %6 %7 %8 %9!
$ERror N!                                                 set up model
$Weight PW_!
$FActor FAC1_ %Z6 FAC2_ %Z6!
$WHile %Z1 ITMH!                                               iterate
$CAlculate UN_=%GL(%SL,1)!                      calculate unit numbers
: %DV=2*%CU(RES_*%LOG(RES_/FIT_))!                  calculate deviance
: %DF=%Z6-1!                                           calculate d.f.
: RES_=(%YV-FIT_)/%SQR(FIT_)!                      calculate residuals
$OUtput %POC!                                          restart output
$TRanscript F H I O W!
$PRint'Marginal Homogeneity Model' :!                      print model
: 'scaled deviance ='%DV' at cycle '*-2 %Z5!
: '        d.f. = '*-2 %DF :!
$SWitch %Z4 PRCY!
$Display E!
$PRint '    unit    observed    fitted  residual'!
$Look (S=-1) UN_ %YV FIT_ RES_!
$$Endmac!
!
$Macro ITMH !                                        iterative fitting
$CAlculate PW_=1/FIT_!
```

```
: %Z5=%Z5+1!
: %Z2=%Z6-1!
$WHile %Z2 MARG!
$Fit C1_+C2_+C3_+C4_+C5_+C6_+C7_+C8_+C9_-1!
$CAlculate FIT_=RES_-%FV!
: %Z3=%DV-%Z2!
: %Z2=%DV!
: %Z1=%IF(%Z4=(%Z3*%Z3>.0001),%Z1-1,0)!          test for convergence
$$Endmac!
!
$Macro MARG !                                       calculate vectors
$CAlculate %%Z2=((FAC1_==%Z2)-(FAC2_==%Z2))*FIT_!
: %Z2=%Z2-1!
$$Endmac!
!
$Macro PRCY !                        message to print if no convergence
$PRint '    (no convergence yet)' :!
$$Endmac!
```

A.4 Non-linear response surface models

```
!  The macro RS3 fits a power transformed three-dimensional response
!surface for any standard GLIM distribution and link.
!  Define the response variable and the distribution. Then, type
!$Use RS3 with the three explanatory variables.
!  The three power parameters are contained in the scalars, %A, %B,
!and %C. These can be used to supply starting values.
!  Macros used: RS3, IRS3
!
$Macro (Local=DERA_,DERB_,DERC_,LOGA_,LOGB_,LOGC_ Argument=V1,V2,V3)
  RS3 !
$OUtput!                     stop output during iterative calculations
$TRanscript!
$DElete DERA_ DERB_ DERC_ LOGA_ LOGB_ LOGC_!
$CA DERA_=DERB_=DERC_=0!
: %A=%A+(%A==0)!                                initial parameter values
: %B=%B+(%B==0)!
: %C=%C+(%C==0)!
: %Z9=20!
: LOGA_=%LOG(V1)!
: LOGB_=%LOG(V2)!
: LOGC_=%LOG(V3)!
: A_=V1!                                         calculate new variables
: B_=V2!
: C_=V3!
: A2_=A_*A_!
: B2_=B_*B_!
: C2_=C_*C_!
```

```
: AB_=A_*B_!
: AC_=A_*C_!
: BC_=B_*C_!
$Fit A_+B_+C_+A2_+B2_+C2_+AB_+AC_+BC_+DERA_+DERB_+DERC_!
$Argument IRS3 V1 V2 V3 LOGA_ LOGB_ LOGC_ DERA_ DERB_ DERC_!
$WHile %Z9 IRS3!                                iterate to solution
$OUtput %POC!                                        restart output
$TRanscript F H I O W!
$Fit A_+B_+C_+A2_+B2_+C2_+AB_+AC_+BC_!          display final model
$PR 'Power parameters: a='%A' b='%B' c='%C :!
$Display E!
$$Endmac!
!
$Macro (Argument=V1,V2,V3,LOGA_,LOGB_,LOGC_,DERA_,DERB_,DERC_) IRS3!
!                    macro to calculate power parameters by iteration
$EXTract %PE!
$CAlculate %A=%A+%PE(11)!                     update power parameters
: %B=%B+%PE(12)!
: %C=%C+%PE(13)!
: A_=V1**%A!                        recalculate transformed variables
: B_=V2**%B!
: C_=V3**%C!
: A2_=A_*A_!
: B2_=B_*B_!
: C2_=C_*C_!
: AB_=A_*B_!
: AC_=A_*C_!
: BC_=B_*C_!
: %Z9=%IF((%PE(11)*%PE(11)<.00001)&(%PE(12)*%PE(12)<.00001)!
  &(%PE(13)*%PE(13)<.00001)&(%Z9/=20),0,%Z9-1)!convergence criterion
$Fit A_+B_+C_+A2_+B2_+C2_+AB_+AC_+BC_!                      fit model
$EXTract %PE!                                   calculate derivatives
$CAlculate DERA_=LOGA_*(%PE(2)*A_+2*%PE(5)*A_*A_+%PE(8)*A_*B_!
  +%PE(9)*A_*C_)!
: DERB_=(%PE(3)*B_+2*%PE(6)*B_*B_+%PE(8)*A_*B_+%PE(10)*B_*C_)*LOGB_!
: DERC_=(%PE(4)*C_+2*%PE(7)*C_*B_+%PE(9)*A_*C_+%PE(10)*B_*C_)*LOGC_!
$Fit A_+B_+C_+A2_+B2_+C2_+AB_+AC_+BC_+DERA_+DERB_+DERC_!
$$Endmac!
```

A.5 Overdispersion models

```
!  The macro EXTB iteratively determines the extra-binomial
!variation of a logistic model (Williams, model II).
!  Prepare and $Fit the desired linear model as usual with
!$ERror B N, then, if Chi-square is much larger than d.f., type
!$Use EXTB.
!  See Williams, D.A. (1982) "Extra-binomial variation in logistic
!linear models." Applied Statistics 31: 144-148.
```

```
!  Macros used: EXTB, EXB1, PRCY, DEFW

$Macro (Local=H_) EXTB !
!$OUtput!
$TRanscript!
$DElete PW_ H_!
$CAlculate %Z1=10!                                    number of iterations
: %Z2=0!
: PW_=1!                                            calculate prior weight
$SWitch %PWF DEFW!
$CAlculate PWT_=PW_!
$Weight PW_!
$Argument EXB1 H_ PW_
$WHile %Z1 EXB1!                                                   iterate
$OUtput %POC!
$TRanscript F H I O W!                                       display model
$SWitch %Z3 PRCY!
$Fit .!
$PRint 'Phi = ' %Z4 :!
$Display E!
$Weight PWT_!                                 reestablish original weight
$$Endmac!
!
$Macro (Argument=H_) EXB1 !            iteratively calculate weights
$Fit .!
$EXTract %VL %WT!
$CAlculate %Z2=%Z2+1!
: %Z3=%DV!
: H_=%PW*%WT*%VL!
: %Z4=%CU(%PW*(1-H_))!
: %Z5=%CU((%BD-1)*%PW*(1-H_))!
: %Z4=(%X2-%Z4)/%Z5!                  calculate dispersion coefficient
: PW_=PWT_/(1+%Z4*(%BD-1))!                        test for convergence
$CAlculate %Z3=(%Z3-%DV)/%DV!
: %Z1=%IF(%Z3=%Z3*%Z3>.0001,%Z1-1,0)!                         convergence?
$$Endmac!
!
$Macro PRCY !                         message to print if no convergence
$PRint '    (no convergence yet)' :!
$$Endmac
!
$Macro DEFW !                                          obtain prior weight
$CAlculate PW_=%PW!
$$Endmac!

!  The macro EXTP iteratively determines the extra-Poisson variation
!of a log linear model.
!  Prepare & $Fit the desired linear model as usual with $ERror P,
```

```
!then, if Chi-square is much larger than d.f., type $Use EXTP.
!  See Breslow, N.E. (1984) "Extra-Poisson variation in log-linear
!models." Applied Statistics 33: 38-44.
!  Macros used: EXTP, EXP2, SIG2, PRCY, DEFW

$Macro EXTP !
$OUtput!
$TRanscript!
$DElete PW_ PWT_!
$CAlculate %Z4=10!                                 number of iterations
: %Z2=0!
: PW_=1!                                         calculate prior weight
$SWitch %PWF DEFW!
$CAlculate PWT_=PW_!
$Fit .!
$EXTract %VL!
$CAlculate %Z5=%CU(%FV*(1-%FV*%VL))!         initial variance estimate
: %Z5=(%X2-%DF)/%Z5!
: %Z3=0!
$Weight PW_!
$WHile %Z4 EXP2!                                                iterate
$OUtput %POC!
$TRanscript F H I O W!                                    display model
$SWitch %Z1 PRCY!
$Fit .!
$Display E!
$PRint 'Variance =' %Z5 :!
$Weight PWT_!                               reestablish original weight
$$Endmac!
!
$Macro SIG2 !                   iteratively calculate variance estimate
$CAlculate %Z6=%Z5!
: %Z5=%CU((%YV-%FV)**2/(%FV*(%FV+1/%Z6)))/%DF!
: %Z1=(%Z5-%Z6)/%Z5!
: %Z1=%Z1*%Z1>.0000001!                            test for convergence
$$Endmac!
!
$Macro EXP2 !                           iteratively calculate weights
$CAlculate %Z2=%Z2+1!
: %Z1=1!
$WHile %Z1 SIG2!
$CAlculate PW_=PWT_/(1+%Z5*%FV)!                    update prior weight
$Fit .!
$RECycle!
$CAlculate %Z3=(%Z3-%DV)/%DV!
: %Z4=%IF(%Z1=%Z3*%Z3>.0001,%Z4-1,0)!              test for convergence
: %Z3=%DV!
$$Endmac
```

```
!
$Macro PRCY !                      message to print if no convergence
$PRint '    (no convergence yet)' :!
$$Endmac!
!
$Macro DEFW !                                     obtain prior weight
$CAlculate PW_=%PW!
$$Endmac!
```

A.6 Random effects models

```
!  The macro RERM fits a random effects variance components model
!to the data for the linear model supplied. The repeated measures
!variable must vary most quickly in the data table.
!  Set up a linear model with $SLength, $FActor, and $ERror. Then,
!define the linear model in a macro called LMOD. For example:
!$Macro LMOD X1+X2+X1.X2 $Endmac. Type $CAlculate %R=the number of
!levels of the random effects variable (set this to one for a
!compound distribution/overdispersion model).
!  Then, type $Use RERM followed by the name of the response
!variable.
!  To define quadrature points for the random effect error
!distribution, create variables QP_, with the points, and QW_ with
!the corresponding weights, and the scalar, %Q, containing the
!number of points. By default, if %Q=0, three quadrature points for
!a normal distribution are used.
!  If underflow is suspected, set %Z to some positive value and try
!again.
!  To try other models, repeat the above steps.
!  See Hinde, J. (1982) "Compound Poisson regression models." in
!Gilchrist, R. (ed.) GLIM 82. Springer Verlag. pp. 109-121.
!  Macros used: RERM, ITEP, FIT, QUAD, NORM, POIS, BINO, GAMM,
!NOR1, POI1, BIN1, GAM1, NOR2, POI2, BIN2, GAM2, BIN3, LMOD

$Macro (Argument=Y Local=Y_,I_,J_,K_,LP_,PP_,PR_,QWW_,PW_,SD_,NN_)
  RERM !
$DElete Y_ I_ J_ K_ LP_ PR_ QWW_ PW_ SD_ NN_!
$Argument ITEP Y PW_ PR_ QWW_ SD_ I_ K_!
: POIS Y!
: BINO Y!
: NOR1 Y_ PW_!
: POI1 Y_ PW_!
: BIN1 Y_ PW_!
: GAM1 Y_ PW_!
: POI2 Y PR_ QWW_ I_ K_!
: BIN2 Y NN_ PR_ QWW_ I_ K_!
: BIN3 NN_ I_!
$OUtput!                                                stop output
```

```
$TRanscript!
$CAlculate %Z9=%ERR!                                   store distribution type
: %Z6=(%Q==0)!                     create quadrature points, if necessary
$SWitch %Z6 QUAD!
$CAlculate NN_=0!                          initialize binomial denominator
: %Z6=%SL*%Q!                                   size of expanded vectors
: %Z7=%SL!                                                store old size
: %Z2=%Z6/%R!
$Variate %Z6 Y_ SD_ QWW_ PW_ LP_ I_ J_ K_!
$CAlculate I_=%GL(%SL,1)!            index for expanding observations
: J_=%GL(%Q,%SL)!               index for expanding quadrature points
: K_=%GL(%Z2,%R)!
: Y_=Y(I_)!                                                       expand
: SD_=QP_(J_)!
: QWW_=QW_(J_)!
$Yvariate Y!
$CYcle!
$Weight!
$Fit #LMOD!                                                  initial fit
$CAlculate LP_=%LP(I_)!                          expand linear predictor
: %Z4=%DV!
: %Z1=%ERR==3!                 check if binomial and set up denominator
$SWitch %Z1 BIN3!
$SLength %Z6!                                               set up model
$SWitch %Z9 NORM POIS BINO GAMM!
$Yvariate Y_!
$RECycle!
$Weight PW_!
$CAlculate %Z3=20!
: %Z5=0!
: %LP=LP_+SD_!                     initial estimates for linear predictor
$WHile %Z3 ITEP!                                    iterate to solution
$OUtput %POC!
$TRanscript F H I O W!
$SWitch %Z9 NOR2 POI2 BIN2 GAM2!                         correct deviance
$EXTract %PE!                                calculate d.f. and deviance
$CAlculate %Z3=%CU(%PE==0)!
: %DF=%Z7-%PL+%Z3!
: %DV=-2*(%Z4+%Z1)!
$PRint 'Random Effects Model' :!                           print results
: 'scaled deviance = '*4 %DV' at cycle'*-2 %Z5!
: '            d.f. = '*-2 %DF :!
$Display E!
$OUtput!
$TRanscript!
$SLength %Z7!                                 reestablish original model
$ASsign N_=NN_!
```

```
$Yvariate Y!
$SWitch %Z9 NORM POIS BINO GAMM!
$OUtput %POC!
$TRanscript F H I O W!
$$Endmac!
!
$Macro QUAD !                             macro for 3 point quadrature
$CAlculate %Q=3!
$ASsign QP_=1.732,.0,-1.732!
: QW_=.1667,.6666,.1667!
$$Endmac!
!
$Macro NORM !                            macros for normal distribution
$ERror N!
$Endmac!
!
$Macro (Argument=Y_,PW_) NOR1 !
$CAlculate %FV=%LP!                                  initial fitted values
: PW_=(Y_-%FV)**2!
: %SC=%CU(PW_)!
: %SC=%SC/%DF!
: PW_=-PW_/2/%SC!                          factor in likelihood function
$$Endmac!
!
$Macro NOR2 !                   macro for correction to normal deviance
$CAlculate %Z1=0!
$PRint : 'Normal Distribution' :!
$$Endmac!
!
$Macro POIS !                           macros for Poisson distribution
$ERror P!
$Endmac!
!
$Macro (Argument=Y_,PW_) POI1 !
$CAlculate %FV=%EXP(%LP)!                            initial fitted values
: PW_=-%FV+Y_*%LP!                         factor in likelihood function
$$Endmac!
!
$Macro (Argument=Y,PR_,QWW_,I_,K_ Local=PP_) POI2 !
!                              macro for correction to Poisson deviance
$WArning!
$CAlculate PP_=-Y(I_)+Y(I_)*%LOG(Y(I_))!
$Use MULT PP_ PR_ QWW_ I_ K_!
$CAlculate %Z1=-%CU(%LOG(PR_)-%Z)!
$WArning!
$PRint : 'Poisson Distribution' :!
$$Endmac!
```

```
!
$Macro BINO !                           macros for binomial distribution
$ERror B N_!
$Endmac!
!
$Macro (Argument=Y_,PW_) BIN1 !
$CAlculate PW_=%EXP(%LP)!
: PW_=PW_/(1+PW_)!
: %FV=%BD*PW_!                                    initial fitted values
: PW_=Y_*%LOG(PW_)+(%BD-Y_)*%LOG(1-PW_)! factor in likelihood funct.
$$Endmac!
!
$Macro (Argument=Y,NN_,PR_,QWW_,I_,K_ Local=PP_) BIN2 !
!                              macro for correction to binomial deviance
$WArning!
$CAlculate PP_=Y(I_)*%LOG(Y(I_)/NN_(I_))!
  +(NN_(I_)-Y(I_))*%LOG(1-Y(I_)/NN_(I_))!
$Use MULT PP_ PR_ QWW_ I_ K_!
$CAlculate %Z1=-%CU(%LOG(PR_)-%Z)!
$WArning!
$PRint : 'Binomial Distribution' :!
$$Endmac!
!
$Macro (Argument=NN_,I_) BIN3 ! macro to set up binomial denominator
$CAlculate NN_=%BD!                              save old denominator
$DElete N_!
$Variate %Z6 N_!
$CAlculate N_=NN_(I_)!                               expand denominator
$$Endmac!
!
$Macro GAMM !                              macros for gamma distribution
$ERror G!
$$Endmac!
!
$Macro (Argument=Y_,PW_) GAM1 !
$CAlculate %FV=1/%LP!                             initial fitted values
: PW_=%DF*(%LOG(Y_/%FV)-Y_/%FV)/%X2!   factor in likelihood function
$$Endmac!
!
$Macro GAM2 !                      macro for correction to gamma deviance
$CAlculate %Z1=0!
$PRint : 'Gamma Distribution' :!
$$Endmac!
!
$Macro (Argument=Y,PW_,PR_,QWW_,SD_,I_,K_) ITEP !
!                                           macro for iterative solution
$CAlculate %Z5=%Z5+1!                                      count cycles
```

```
: %Z8=%Z4!                                          keep old likelihood
$SWitch %Z9 NOR1 POI1 BIN1 GAM1!
$Use MULT PW_ PR_ QWW_ I_ K_!
$CAlculate PW_=PW_/PR_(I_)!
$Fit (#LMOD)(I_)+SD_!
$CAlculate %Z4=%CU(%LOG(PR_)-%Z)!                  test for convergence
: %Z3=%IF(((%Z8-%Z4)**2)>.001,%Z3-1,0)!
$$Endmac!
!
$Macro (Argument=PP_,PR_,QWW_,I_,K_ Local=PPR_) MULT !
!                           macro to multiply probabilities together
$DElete PPR_!
$Variate %Z2 PPR_!
$CAlculate PPR_=0!
: PPR_(K_)=PPR_(K_)+PP_!
: PP_=QWW_*%EXP(PPR_(K_)+%Z)!
$Variate %Z7 PR_!
$CAlculate PR_=0!       add up probabilities to calculate new weight
: PR_(I_)=PR_(I_)+PP_!
$$Endmac!
```

Bibliography

1. Aalen, O.O. (1978) Nonparametric inference for a family of counting processes. *Annals of Statistics* **6**, 701–726.
2. Aalen, O.O. (1989) A linear regression model for the analysis of life times. *Statistics in Medicine* **8**, 907–925.
3. Agresti, A. (1984) *Analysis of Ordinal Categorical Data.* New York: Wiley.
4. Agresti, A. (1990) *Categorical Data Analysis.* New York: John Wiley.
5. Aickin, M. (1983) *Linear Statistical Analysis of Discrete Data.* New York: Wiley.
6. Aitkin, M. (1994) Model choices in single samples from the exponential and double exponential families using the posterior Bayes factor. *Statistics and Computing* **4**, (in press).
7. Aitkin, M., Anderson, D., Francis, B., and Hinde, J. (1989) *Statistical Modelling in GLIM.* Oxford: Oxford University Press.
8. Aiuppa, T.A. (1988) Evaluation of Pearson curves as an approximation of the maximum probable annual aggregate loss. *Journal of Risk and Insurance* **55**, 425–441.
9. Andersen, E.B. (1977) Multiplicative Poisson models with unequal cell rates. *Scandinavian Journal of Statistics* **4**, 153–158.
10. Andersen, E.B. (1980) *Discrete Statistical Models with Social Science Applications.* Amsterdam: North Holland.
11. Andersen, E.B. (1991) *Statistical Analysis of Categorical Data.* Berlin: Springer Verlag.
12. Andersen, P.K. and Borgan, Ø. (1985) Counting process models for life history data: a review. *Scandinavian Journal of Statistics* **12**, 97–158.
13. Anderson, J.A. (1984) Regression and ordered categorical variables. *Journal of the Royal Statistical Society* **B46**, 1–30.
14. Andrews, D.F. and Herzberg, A.M. (1985) *Data. A Collection of Problems from Many Fields for the Student and Research Worker.* Berlin: Springer Verlag.
15. Arnold, B.C. and Strauss, D.J. (1991) Bivariate distributions with conditionals in prescribed exponential families. *Journal of the Royal Statistical Society* **B53**, 365–375.
16. Ashford, J.A. (1959) An approach to the analysis of data for semi-quantal responses in biological assay. *Biometrics* **15**, 573–581.
17. Ashford, J.R. and Sowden, R.R. (1970) Multivariate probit analysis. *Biometrics* **26**, 535–546.

18. Barnett, V. and Lewis, T. (1984) *Outliers in Statistical Data.* New York: John Wiley.

19. Beitler, P.J. and Landis, J.R. (1985) A mixed-effects model for categorical data. *Biometrics* **41**, 991–1000.

20. Bishop, Y.M.M. and Fienberg, S.E. (1969) Incomplete two-dimensional contingency tables. *Biometrics* **25**, 119–123.

21. Bishop, Y.M.M., Fienberg, S.E., and Holland, P.W. (1975) *Discrete Multivariate Analysis: Theory and Practice.* Cambridge: MIT Press.

22. Bissell, A.F. (1972) A negative binomial model with varying element sizes. *Biometrika* **59**, 435–441.

23. Borgan, Ø. (1984) Maximum likelihood estimation in parametric counting process models, with applications to censored failure time data. *Scandinavian Journal of Statistics* **11**, 1–16.

24. Brass, W. (1959) Simplified methods of fitting the truncated negative binomial distribution. *Biometrika* **45**, 59–68.

25. Breslow, N.E. (1982) Covariance adjustment of relative-risk estimates in matched studies. *Biometrics* **38**, 661–672.

26. Breslow, N.E. (1984) Extra-Poisson variation in log-linear models. *Journal of the Royal Statistical Society* **C33**, 38–44.

27. Breslow, N.E. and Day, N.E. (1982) *Statistical Methods in Cancer Research.* Vol. 1. *The Analysis of Case-control studies.* Lyon: International Agency for Research on Cancer.

28. Brown, P.J., Stone, J., and Ord-Smith, C. (1983) Toxaemic signs during pregnancy. *Journal of the Royal Statistical Society* **C32**, 69–72.

29. Burridge, J. (1981) Empirical Bayes analysis of survival time data. *Journal of the Royal Statistical Society* **B43**, 65–75.

30. Chatfield, C., Ehrenberg, A.S.C., and Goodhardt, G.J. (1966) Progress on a simplified model of stationary purchasing behaviour. *Journal of the Royal Statistical Society* **B28**, 317–367.

31. Cheke, R.A. (1985) Winter feeding assemblies, wing lengths and weights of British dunnocks. In Morgan, B.J.T. and North, P.M. (eds.) *Statistics in Ornithology.* Berlin: Springer Verlag, pp. 13–24.

32. Christensen, R. (1990) *Log-Linear Models.* Berlin: Springer Verlag.

33. Cohen, J.E. (1976) The distribution of the chi-squared statistic under cluster sampling from contingency tables. *Journal of the American Statistical Association* **71**, 665–670.

34. Coleman, J.S. (1964) *Introduction to Mathematical Sociology.* Glencoe: The Free Press.

35. Collett, D. (1991) *Modelling Binary Data.* London: Chapman and Hall.

36. Conaway, M.R. (1989) Analysis of repeated categorical measurements with conditional likelihood methods. *Journal of the American Statistical Association* **84**, 53–62.

37. Cormack, R.M. (1985) Examples of the use of GLIM to analyse capture–recapture studies. In Morgan, B.J.T. and North, P.M. (eds.) *Statistics in Ornithology.* Berlin: Springer Verlag, pp. 243–273.

38. Cormack, R.M. (1989) Log-linear models for capture–recapture. *Biometrics* **45**, 395–413.
39. Cox, D.R. (1966) A simple example of a comparison involving quantal data. *Biometrika* **53**, 215–220.
40. Cox, D.R. (1970) *Analysis of Binary Data.* London: Methuen.
41. Cox, D.R. and Lewis, P.A.W. (1966) *The Statistical Analysis of Series of Events.* London: Methuen.
42. Cox, D.R. and Snell, E.J. (1989) *The Analysis of Binary Data.* London: Chapman and Hall.
43. Crouchley, R., Davies, R.B., and Pickles, A.R. (1982) Identification of some recurrent choice processes. *Journal of Mathematical Sociology* **9**, 63–73.
44. Crowder, M.J. (1978) Beta-binomial Anova for proportions. *Journal of the Royal Statistical Society* **C27**, 34–37.
45. Dahiya, R.C. and Gross, A.J. (1973) Estimating the zero class from a truncated Poisson sample. *Journal of the American Statistical Association* **68**, 731–733.
46. Davidson, R.R. (1970) On extending the Bradley–Terry model to accommodate ties in paired comparison experiments. *Journal of the American Statistical Association* **65**, 317–328.
47. de Angelis, D. and Gilks, W.R. (1994) Estimating acquired immune deficiency syndrome incidence accounting for reporting delay. *Journal of the Royal Statistical Society* **A157**, 31–40.
48. Decarli, A., Francis, B., Gilchrist, R., and Seeber, G.U.H. (1989) *Statistical Modelling.* Berlin: Springer Verlag.
49. de Jong, P. and Greig, M. (1985) Models and methods for pairing data. *Canadian Journal of Statistics* **13**, 233–241.
50. Derman, C., Gleser, L.J., and Olkin, I. (1973) *A Guide to Probability Theory and Application.* New York: Holt, Rinehart, and Winston.
51. Dobson, A.J. (1990) *An Introduction to Generalized Linear Models.* London: Chapman and Hall.
52. Duncan, O.D. (1979) How destination depends on origin in the occupational mobility table. *American Journal of Sociology* **84**, 793–803.
53. Durkheim, E. (1897) *Le Suicide: Etude de Sociologie.* Paris: Presses universitaires de France.
54. Edwards, D.E. and Havranek, T. (1985) A fast procedure for model search in multidimensional contingency tables. *Biometrika* **72**, 339–351.
55. Efron, B. (1986) Double exponential families and their use in generalized linear regression. *Journal of the American Statistical Association* **81**, 709–721.
56. Everitt, B.S. (1977) *The Analysis of Contingency Tables.* London: Chapman and Hall.
57. Farewell, V.T. (1982) A note on regression analysis of ordinal data with variability of classification. *Biometrika* **69**, 533–538.
58. Fienberg, S.E. (1977) *The Analysis of Cross-Classified Categorical Data.* Cambridge: MIT Press.

59. Fingleton, B. (1984) *Models of Category Counts.* Cambridge: Cambridge University Press.

60. Finney, D.J. (1978) *Statistical Method in Biological Assay.* London: C. Griffin.

61. Finney, D.J. and Varley, G.C. (1955) An example of the truncated Poisson distribution. *Biometrics* **11**, 387–391.

62. Fisher, R.A. (1958) *Statistical Methods for Research Workers.* Edinburgh: Oliver and Boyd.

63. Fleming, T.R. and Harrington, D.P. (1991) *Counting Processes and Survival Analysis.* New York: John Wiley.

64. Francis, B., Green, M., and Payne, C. (1993) *The GLIM System. Release 4 Manual.* Oxford: Oxford University Press.

65. Francom, S.F., Chuang-Stein, C., and Landis, J.R. (1989) A log-linear model for ordinal data to characterize differential change among treatments. *Statistics in Medicine* **8**, 571–582.

66. Gelfand, A.E. and Dalal, S.R. (1990) A note on overdispersed exponential families. *Biometrika* **77**, 55–64.

67. Gilchrist, R. (1981) Calculations of residuals for all GLIM models. *GLIM Newsletter* **4**, 26–27.

68. Gilchrist, R. (1982a) GLIM syntax for adjusted residuals. *GLIM Newsletter* **6**, 64–65.

69. Gilchrist, R. (1982b, ed.) *GLIM82.* Berlin: Springer Verlag.

70. Gilchrist, R., Francis, B., and Whittaker, J. (1985, eds.) *Generalized Linear Models.* Berlin: Springer Verlag.

71. Glenn, W.A. and David, H.A. (1960) Ties in paired-comparison experiments using a modified Thurstone–Mosteller model. *Biometrics* **16**, 86–109.

72. Goodman, L.A. (1962) Statistical methods for analyzing processes of change. *American Journal of Sociology* **68**, 57–78.

73. Goodman, L.A. (1979) Simple models for the analysis of association in cross-classifications having ordered categories. *Journal of the American Statistical Association* **74**, 537–552.

74. Goodman, L.A. (1981) Association models and canonical correlation in the analysis of cross-classifications having ordered categories. *Journal of the American Statistical Association* **76**, 320–334.

75. Gottschau, A. (1994) Markov chain models for multivariate binary panel data. *Scandinavian Journal of Statistics* **21**, 57–71.

76. Greenland, S. (1994) Alternative models for ordinal logistic regression. *Statistics in Medicine* **13**, 1665–1677.

77. Grizzle, J.E., Starmer, C.F., and Koch, G.G. (1969) Analysis of categorical data by linear models. *Biometrics* **25**, 489–504.

78. Haberman, S.J. (1974a) *The Analysis of Frequency Data.* Chicago: University of Chicago Press.

79. Haberman, S.J. (1974b) Log-linear models for frequency data with ordered classifications. *Biometrics* **30**, 589–600.

80. Haberman, S.J. (1978) *Analysis of Qualitative Data.* Vol. I. *Introductory Topics.* New York: Academic Press.

81. Haberman, S.J. (1979) *Analysis of Qualitative Data.* Vol. II. *New Developments.* New York: Academic Press.

82. Hagenaars, J.A. (1990) *Categorical Longitudinal Data. Log-Linear Panel, Trend, and Cohort Analysis.* Newbury Park: Sage.

83. Han, A. and Hausman, J.A. (1990) Flexible parametric estimation of duration and competing risk models. *Journal of Applied Econometrics* **5**, 1–28.

84. Härdle, W. and Stoker, T.M. (1989) Investigating smooth multiple regression by the method of average derivatives. *Journal of the American Statistical Association* **84**, 986–995.

85. Hay, J.W. and Wolak, F.A. (1994) A procedure for estimating the unconditional cumulative incidence curve and its variability for the human immunodeficiency virus. *Journal of the Royal Statistical Society* **C43**, 599–624.

86. Healy, M.J.R. (1988) *Glim: An Introduction.* Oxford: Oxford University Press.

87. Heckman, J.J. and Willis, R.J. (1977) A beta-logistic model for the analysis of sequential labor force participation by married women. *Journal of Political Economy* **85**, 27–58.

88. Higgins, J.E. and Koch, G.G. (1977) Variable selection and generalized chi-square analysis of categorical data applied to a large cross-sectional occupational health survey. *International Statistical Review* **45**, 51–62.

89. Hinde, J. (1982) Compound Poisson regression models. in Gilchrist, R. (ed.) *GLIM 82.* Berlin: Springer Verlag, pp. 109–121.

90. Hinkley, D.V., Reid, N., and Snell, E.J. (1990) *Statistical Theory and Modelling.* London: Chapman and Hall.

91. Holtbrügge, W. and Schumacher, M. (1991) A comparison of regression models for the analysis of ordered categorical data. *Journal of the Royal Statistical Society* **C40**, 249–259.

92. Hosmer, D.W. and Lemeshow, S. (1989) *Applied Logistic Regression.* New York: John Wiley.

93. Hutchison, D. (1985) Ordinal variable regression using the McCullagh (proportional odds) model. *GLIM Newsletter* **9**, 9–17.

94. Irwin, J.O. (1975) The generalized Waring distribution. Part II. *Journal of the Royal Statistical Society* **A138**, 204–284.

95. Jansen, J. (1990) On the statistical analysis of ordinal data when extravariation is present. *Journal of the Royal Statistical Society* **C39**, 75–84.

96. Jarrett, R.G. (1979) A note on the intervals between coal-mining disasters. *Biometrika* **66**, 191–193.

97. Kalbfleisch, J.G. (1985) *Probability and Statistical Inference.* Vol. 2. Berlin: Springer Verlag.

98. Kitagawa, G. (1987) Non-Gaussian state-space modeling of nonstationary time series. *Journal of the American Statistical Association* **82**, 1032–1063.

99. Klotz, J. (1973) Statistical inference in Bernoulli trials with dependence. *Annals of Statistics* **1**, 373–379.

100. Knoke, D. and Burke, P.J. (1980) *Log-Linear Models.* Beverley Hills: Sage.

101. Koch, G.G., Landis, J.R., Freeman, J.L., Freeman, D.H., and Lehnen, R.G. (1977) A general methodology for the analysis of experiments with repeated measurement of categorical data. *Biometrics* **33**, 133–158.

102. Kocherlakota, S. and Kocherlakota, K. (1990) Tests of hypotheses for the weighted binomial distribution. *Biometrics* **46**, 645–656.

103. Landis, J.R. and Koch, G.G. (1977) The measurement of observer agreement for categorical data. *Biometrics* **33**, 159–174.

104. Lawal, H.B. and Upton, G.J.G. (1990) Alternative interaction structures in square contingency tables having ordered classificatory variables. *Quality and Quantity* **24**, 107–127.

105. Lee, E.T. (1992) *Statistical Methods for Survival Data Analysis.* New York: John Wiley.

106. Leiter, R.E. and Hamdan, M.A. (1973) Some bivariate probability models applicable to traffic accidents and fatalities. *International Statistical Review* **41**, 87–100.

107. Lindsey, J.K. (1973) *Inferences from Sociological Survey Data: A Unified Approach.* Amsterdam: Elsevier.

108. Lindsey, J.K. (1974a) Comparison of probability distributions. *Journal of the Royal Statistical Society* **B36**, 38–47.

109. Lindsey, J.K. (1974b) Construction and comparison of statistical models. *Journal of the Royal Statistical Society* **B36**, 418–425.

110. Lindsey, J.K. (1975) Likelihood analysis and test for binary data. *Journal of the Royal Statistical Society* **C24**, 1–16.

111. Lindsey, J.K. (1989) *The Analysis of Categorical Data Using GLIM.* Berlin: Springer Verlag.

112. Lindsey, J.K. (1992) *The Analysis of Stochastic Processes Using GLIM.* Berlin: Springer Verlag.

113. Lindsey, J.K. (1993) *Models for Repeated Measurements.* Oxford: Oxford University Press.

114. Lindsey, J.K. (1995a) *Introductory Statistics: The Modelling Approach.* Oxford: Oxford University Press.

115. Lindsey, J.K. (1995b) Fitting parametric counting processes by using log linear models. *Journal of the Royal Statistical Society* **C44**, (in press).

116. Lindsey, J.K., Alderdice, D.F., and Pienaar, L.V. (1970) Analysis of nonlinear models — the nonlinear response surface. *Journal of the Fisheries Research Board of Canada* **27**, 765–791.

117. Lindsey, J.K., Jones, B., and Lewis, J.A. (1995) Analysis of cross-over trials for duration data. *Statistics in Medicine* **14**, (in press).

118. Lindsey, J.K. and Mersch, G. (1992) Fitting and comparing probability distributions with log linear models. *Computational Statistics and Data Analysis* **13**, 373–384.

119. Lombard, H.L. and Doering, C.R. (1947) Treatment of the four-fold table by partial association and partial correlation as it relates to public health problems. *Biometrics* **3**, 123–128.

120. Maxwell, A.E. (1961) *Analysing Qualitative Data.* London: Methuen.

121. McCullagh, P. and Nelder, J.A. (1989) *Generalized Linear Models.* London: Chapman and Hall.

122. Mislevy, R. (1985) Estimation of latent group effects. *Journal of the American Statistical Association* **80**, 993–997.

123. Morgan, B.J.T. (1992) *Analysis of Quantal Response Data.* London: Chapman and Hall.

124. Nelder, J.A. (1974) Log linear models for contingency tables: a generalization of classical least-squares. *Journal of the Royal Statistical Society* **C23**, 323–329.

125. Nelder, J.A. and Wedderburn, R.W.M. (1972) Generalized linear models. *Journal of the Royal Statistical Society* **A135**, 370–384.

126. Pegram, G.G.S. (1980) An autoregressive model for multilag Markov chains. *Journal of Applied Probability* **17**, 350–362.

127. Piegorsch, W.W. (1992) Complementary log regression for generalized linear models. *American Statistician* **46**, 94–99.

128. Plackett, R.L. (1965)A class of bivariate distributions. *Journal of the American Statistical Association* **60**, 516–522.

129. Plackett, R.L. (1974) *The Analysis of Categorical Data.* London: Griffin.

130. Pregibon, D. (1981) Logistic regression diagnostics. *Annals of Statistics* **9**, 705–724.

131. Pregibon, D. (1982) Score tests with applications. In Gilchrist (1982b), pp. 87–97.

132. Rasch, G. (1960) *Probabilistic Models for some Intelligence and Attainment Tests.* Copenhagen: Danish Institute for Educational Research.

133. Reynolds, H.Y. (1977) *The Analysis of Cross-Classifications.* New York: Free Press.

134. Rugg, D.J. and Buech, R.R. (1990) Analyzing time budgets with Markov chains. *Biometrics* **46**, 1123–1131.

135. Santner, T.J. and Duffy, D.E. (1989) *The Statistical Analysis of Discrete Data.* Berlin: Springer Verlag.

136. Sewell, W.H. and Shah, V.P. (1968) Social class, parental encouragement, and educational aspirations. *American Journal of Sociology* **73**, 559–572.

137. Skellam, J.G. (1948) A probability distribution derived from the binomial distribution by regarding the probability of success as variable between sets of trials. *Journal of the Royal Statistical Society* **B10**, 257–261.

138. Sokal, R.R. and Rohlf, F.J. (1969) *Biometry. The Principles and Practice of Statistics in Biological Research.* San Francisco: W.H. Freeman.

139. Sprent, P. (1993) *Applied Nonparametric Statistical Methods.* London: Chapman and Hall.

140. Stouffer, S.A. and Toby, J. (1951) Role conflict and personality. *American Journal of Sociology* **56**, 395–406.

141. Stuart, A. (1955) A test for homogeneity of the marginal distribution in a two-way classification. *Biometrika* **42**, 412–416.

142. Stukel, T.A. (1988) Generalized logistic models. *Journal of the American Statistical Association* **83**, 426–431.

143. Tjur, T. (1982) A connection between Rasch's item analysis model and a multiplicative Poisson model. *Scandinavian Journal of Statistics* **9**, 23–30.

144. Upton, G.J.G. (1978) *The Analysis of Cross-Tabulated Data.* New York: Wiley.

145. Upton, G.J.G. and Fingleton, B. (1989) *Spatial Data Analysis by Example.* Volume 2. *Categorical and Directional Data.* New York: John Wiley.

146. van der Heijden, P.G.M., Jansen, W., Francis, B., and Seeber, G.U.H. (1992) *Statistical Modelling.* Amsterdam: North Holland.

147. van Houwelingen, H.C. and Zwinderman, K.H. (1993) A bivariate approach to meta-analysis. *Statistics in Medicine* **12**, 2273–2284.

148. Vidmar, T.J., McKean, J.W., and Hettmansperger, T.P. (1992) Robust procedures for drug combination problems with quantal responses. *Journal of the Royal Statistical Society* **C41**, 299–315.

149. Wei, L.J. and Lachin, J.M. (1984) Two-sample asymptotically distribution-free tests for incomplete multivariate observations. *Journal of the American Statistical Association* **79**, 653–661.

150. Wermuth, N. (1976) Model search among multiplicative models. *Biometrics* **32**, 253–264.

151. Whittaker, J. (1990) *Graphical Models in Applied Multivariate Statistics.* New York: Wiley.

152. Whittemore, A.S. and Gong, G. (1991) Poisson regression with misclassified counts: application to cervical cancer mortality rates. *Journal of the Royal Statistical Society* **C40**, 81–93.

153. Wilkinson, G.N. and Rogers, C.E. (1973) Symbolic description of factorial models for analysis of variance. *Journal of the Royal Statistical Society* **C22**, 392–399.

154. Williams, D.A. (1982) Extra-binomial variation in logistic linear models. *Journal of the Royal Statistical Society* **C31**, 144–148.

155. Zeger, S.L. (1988) A regression model for time series of counts. *Biometrika* **75**, 621–629.

156. Zelterman, D. (1987) Goodness-of-fit tests for large sparse multinomial distributions. *Journal of the American Statistical Association* **82**, 624–629.

Index